IMPROV
SEISMIC MONITORING
IMPROVED
DECISION-MAKING

Assessing the Value of Reduced Uncertainty

Committee on the Economic Benefits of Improved Seismic Monitoring

Committee on Seismology and Geodynamics

Board on Earth Sciences and Resources

Division on Earth and Life Studies

NATIONAL RESEARCH COUNCIL
OF THE NATIONAL ACADEMIES

THE NATIONAL ACADEMIES PRESS
Washington, D.C.
www.nap.edu

THE NATIONAL ACADEMIES PRESS, 500 Fifth Street, N.W., Washington, DC 20001

NOTICE: The project that is the subject of this report was approved by the Governing Board of the National Research Council, whose members are drawn from the councils of the National Academy of Sciences, the National Academy of Engineering, and the Institute of Medicine. The members of the committee responsible for the report were chosen for their special competences and with regard for appropriate balance.

The views and conclusions contained in this document are those of the authors and should not be interpreted as representing the opinions or policies of the U.S. government. Mention of trade names or commercial products does not constitute their endorsement by the U.S. government. Supported by the U.S. Geological Survey, Department of the Interior, under assistance Award No. 03HQGR0114.

International Standard Book Number 0-309-09695-2 (Book)
International Standard Book Number 0-309-55178-1 (PDF)
Library of Congress Control Number 2005937437

Additional copies of this report are available from the National Academies Press, 500 Fifth Street, N.W., Lockbox 285, Washington, DC 20055; (800) 624-6242 or (202) 334-3313 (in the Washington metropolitan area); Internet http://www.nap.edu

Cover: Design by Michele de la Menardiere.

Printed in the United States of America.

THE NATIONAL ACADEMIES

Advisers to the Nation on Science, Engineering, and Medicine

The **National Academy of Sciences** is a private, nonprofit, self-perpetuating society of distinguished scholars engaged in scientific and engineering research, dedicated to the furtherance of science and technology and to their use for the general welfare. Upon the authority of the charter granted to it by the Congress in 1863, the Academy has a mandate that requires it to advise the federal government on scientific and technical matters. Dr. Ralph J. Cicerone is president of the National Academy of Sciences.

The **National Academy of Engineering** was established in 1964, under the charter of the National Academy of Sciences, as a parallel organization of outstanding engineers. It is autonomous in its administration and in the selection of its members, sharing with the National Academy of Sciences the responsibility for advising the federal government. The National Academy of Engineering also sponsors engineering programs aimed at meeting national needs, encourages education and research, and recognizes the superior achievements of engineers. Dr. Wm. A. Wulf is president of the National Academy of Engineering.

The **Institute of Medicine** was established in 1970 by the National Academy of Sciences to secure the services of eminent members of appropriate professions in the examination of policy matters pertaining to the health of the public. The Institute acts under the responsibility given to the National Academy of Sciences by its congressional charter to be an adviser to the federal government and, upon its own initiative, to identify issues of medical care, research, and education. Dr. Harvey V. Fineberg is president of the Institute of Medicine.

The **National Research Council** was organized by the National Academy of Sciences in 1916 to associate the broad community of science and technology with the Academy's purposes of furthering knowledge and advising the federal government. Functioning in accordance with general policies determined by the Academy, the Council has become the principal operating agency of both the National Academy of Sciences and the National Academy of Engineering in providing services to the government, the public, and the scientific and engineering communities. The Council is administered jointly by both Academies and the Institute of Medicine. Dr. Ralph J. Cicerone and Dr. Wm. A. Wulf are chair and vice chair, respectively, of the National Research Council.

www.national-academies.org

COMMITTEE ON THE ECONOMIC BENEFITS OF IMPROVED SEISMIC MONITORING

CHRIS D. POLAND, *Chair*, Degenkolb Engineers, San Francisco, California
JAMES AMENT, State Farm Fire and Casualty Co., Bloomington, Illinois
DAVID S. BROOKSHIRE, The University of New Mexico, Albuquerque
JAMES D. GOLTZ, California Governor's Office of Emergency Services, Pasadena
PETER GORDON, University of Southern California, Los Angeles
STEPHANIE A. KING, Weidlinger Associates, Inc., Los Altos, California
HOWARD KUNREUTHER, The Wharton School, University of Pennsylvania, Philadelphia
STUART P. NISHENKO, Pacific Gas and Electric Company, San Francisco, California
ADAM Z. ROSE, The Pennsylvania State University, University Park
HOPE A. SELIGSON, ABS Consulting, Irvine, California
PAUL G. SOMERVILLE, URS Group, Inc., Pasadena, California

Liaison from Committee on Seismology and Geodynamics:
TERRY C. WALLACE, Jr., Los Alamos National Laboratory, New Mexico

National Research Council Staff
DAVID A. FEARY, Study Director
JENNIFER T. ESTEP, Administrative Associate
RADHIKA S. CHARI, Senior Project Assistant (until 5/04)
AMANDA M. ROBERTS, Project Assistant (from 7/04)

COMMITTEE ON SEISMOLOGY AND GEODYNAMICS

TERRY C. WALLACE, Jr., *Chair*, Los Alamos National Laboratory, New Mexico
ALAN LEVANDER, *Vice-Chair*, Rice University, Houston, Texas
ROLAND BÜRGMANN, University of California, Berkeley
ADAM M. DZIEWONSKI, Harvard University, Cambridge, Massachusetts
WILLIAM E. HOLT, State University of New York at Stony Brook
LOUISE H. KELLOGG, University of California, Davis
M. MEGHAN MILLER, Central Washington University, Ellensburg
JACK R. MURPHY, Science Applications International Corporation, Arlington, Virginia
PAUL G. SILVER, Carnegie Institution of Washington, D.C.
AARON A. VELASCO, University of Texas at El Paso
RU-SHAN WU, University of California, Santa Cruz

National Research Council Staff
DAVID A. FEARY, Study Director
VERNA J. BOWEN, Administrative Assistant

Preface

For those of us who visit and assess areas devastated by earthquakes and have responsibility for ensuring that the damaging effects of earthquakes are minimized, the value of seismic monitoring as one of the essential tools is absolutely clear and unchallenged. However, providing an economic assessment of the value of this tool is a different and difficult issue, and one that has long challenged the nation's scientists and engineers.

This study, commissioned by the U.S. Geological Survey, is aimed specifically at assessing the economic benefits of modernizing and expanding seismic monitoring activities in the United States, so that the value derived from monitoring data can be compared to other activities competing for the same resources. The National Research Council—in recognition of the multidisciplinary nature of this issue—populated the study committee with representatives from the range of professions involved with geoscience, emergency management, and earthquake engineering issues, together with expert economists to ensure that the benefit analysis was undertaken with appropriate rigor. The committee accepted public testimony, deliberated thoughtfully and with considerable skepticism, and developed this report to clearly set the stage, define the issues, and discuss the costs and benefits that improved seismic monitoring will have on all aspects of earthquake science and engineering.

The committee commenced this study with the expectation that it would collectively be able to identify the many areas where improved seismic monitoring information would contribute to mitigating earthquake losses and be able to use a diverse range of existing information to quantify the economic benefits. In the end, the committee concluded that although it was possible to describe the numerous potential benefits, attempts to quantify them rigorously proved elusive because the required information either does not exist or is not routinely collected. In keeping with its charge, the committee used a range of assumptions to derive a very approximate estimate of potential performance-based engineering benefits to illustrate the complexity of this task as well as the magnitude of potential benefits.

The recent tragedy in nations surrounding the northern Indian Ocean, caused by the 2004 Sumatran earthquake and tsunami, provided vivid testimony to the awesome power of forces within the earth's crust, and the enormous potential that these forces pose for devastating loss of life and economic disruption. This event focused national and international attention on the capabilities of warning systems for mitigating natural disasters, leading to accelerated implementation of long-established plans to expand tsunami warning systems. Will it take a similarly devastating earthquake in the United States to accelerate long-established—but only partially funded—plans to broaden seismic monitoring programs to maximize the potential for earthquake hazard mitigation?

On behalf of the committee, I would like to acknowledge and thank all the scientists and engineers who made presentations at our four committee meetings. I wish to also thank the committee members for their thoughtful, pointed, and candid views and their willingness to listen, discover the benefits, and come to agreement. Most of all, I want to thank David Feary and the other members of the NRC staff for their hard work and diligence in keeping us organized, focused, and understandable.

Chris D. Poland
Chair

Acknowledgments

The committee would like to express its appreciation to the many individuals who provided briefings and other information during the information-gathering process: Richard Bernknopf, Steven Bohlen, Dan Byers, Stephen Cauffman, Bruce Clark, Lloyd Cluff, Richard Eisner, Bill Ellsworth, John Filson, Jason Freihage, Linda Gundersen, Robert Herrmann, Richard Howe, Lucy Jones, Patrick Leahy, William Leith, E.V. Leyendecker, Mike Mahoney, Steven McCabe, Charles Meade, Priscilla Nelson, Bela Palfalvi, Paul Reasenberg, Cliff Roblee, Doug Sandy, Woody Savage, Kaye Shedlock, David Simpson, Zan Turner, Craig Weaver, Gene Whitney, Mitch Withers, and Darryl Young. The committee particularly acknowledges the provision of information from Ron Tognazzini and Craig Davis (Los Angeles Department of Water and Power).

This report has been reviewed in draft form by individuals chosen for their diverse perspectives and technical expertise, in accordance with procedures approved by the National Research Council's Report Review Committee. The purpose of this independent review is to provide candid and critical comments that will assist the institution in making its published report as sound as possible and to ensure that the report meets institutional standards for objectivity, evidence, and responsiveness to the study charge. The review comments and draft manuscript remain confidential to protect the integrity of the deliberative process. We wish to thank the following individuals for their review of this report:

Gail M. Atkinson, Department of Earth Sciences, Carleton University, Ottawa, Canada

Stephanie E. Chang, Institute for Resources, Environment and Sustainability, and School of Community and Regional Planning, University of British Columbia, Vancouver, Canada
Ronald T. Eguchi, ImageCat, Inc., Long Beach, California
Robert M. Hamilton, Zelienople, Pennsylvania
Peter J. May, Political Science Department, University of Washington, Seattle
Claire B. Rubin, Claire B. Rubin & Associates, Arlington, Virginia
Craig Tillman, Wyndham Partners Consulting Ltd., (an affiliate of Renaissance Reinsurance Ltd.), Laguna Niguel, California
Richard J. Zeckhauser, John F. Kennedy School of Government, Harvard University, Cambridge, Massachusetts

Although the reviewers listed above have provided many constructive comments and suggestions, they were not asked to endorse the conclusions or recommendations nor did they see the final draft of the report before its release. The review of this report was overseen by William J. Petak, University of Southern California, Los Angeles. Appointed by the National Research Council, he was responsible for ensuring that an independent examination of this report was carried out in accordance with institutional procedures and that all review comments were carefully considered. Responsibility for the final content of this report rests entirely with the authoring committee and the institution.

Contents

Executive Summary

Protecting people, the built environment, and the nation's economy from the effects of devastating earthquakes has been the focus of scientific and engineering endeavor for more than 100 years. The need for such protection has never been greater—approximately 30 percent of the population (75 million people) and 50 percent of the national building stock ($8.6 trillion) are located in areas prone to damaging earthquakes. Annualized building and buildings-related earthquake losses in the United States are estimated to be more than $5.6 billion per year, with a single significant earthquake having the potential to cause losses of more than $100 billion.

Members of many professions are actively seeking to understand and mitigate the consequences of major earthquakes—earth scientists focus on understanding the source and nature of the strong shaking and defining the hazard, block by block, to our cities; structural engineers and their design professional colleagues work to determine the optimum ways to mitigate damage and endeavor to design their structures appropriately; economists and policy analysts focus on determining the appropriate framework for evaluating the benefits and cost-effectiveness of implementing various mitigation measures, including improved seismic monitoring; insurance professionals and their loss estimation consultants work to determine the expected dollar losses that could occur and provide risk mitigation products to share the cost of damage; and the emergency management community defines scenario events and plans, practices, and assists with recovery from earthquake disasters. All of these professions are users of information derived from seismic monitoring.

The advances made by each of these professions depends on basic scientific research, on applied research founded on a clear understanding of what has happened after each earthquake, on collaboration to determine how best to improve the processes they use, and on education to disseminate best practices to all practitioners in each profession. The significant advances that have been made over the past 50 years have been aided by the availability of seismic monitoring data—records of earthquake events recorded by weak and strong motion instruments. All components of the built environment—buildings, bridges, roads, utility networks, and dams—share in the benefits. The records obtained have contributed to seismic hazard maps, to building codes and loss estimation programs, and to innovative emergency response tools that graphically display the areas subject to greatest damage.

In this process, it has become clear that earthquakes are very complex phenomena, each one leaving a different signature of ground shaking that varies in intensity and characteristics depending on location, ground conditions, type and geometry of faulting, and magnitude. Seismic monitoring holds the key to understanding both the seismic hazard and the response of the built environment so that proper preparations can be made for the future. However, despite the widely appreciated benefits of seismic monitoring data, we currently have in place only a fraction of the modern instruments needed to capture the essence of the earthquakes that are occurring. In fact, over the last 30 years in the United States, a number of opportunities have been missed to record events, yield new insights, and eventually reduce the cost of earthquakes.

The Advanced National Seismic System (ANSS) is envisioned as a state-of-the-art network of seismic monitoring instruments that will provide data on both earthquake occurrence and infrastructure response to earthquake ground shaking. The national benefits of such a system are easy to appreciate from a qualitative perspective; however, the benefits are inherently very difficult to quantify in terms of dollars "saved" or losses avoided. The ANSS was proposed by the U.S. Geological Survey (USGS) in 1999 and was endorsed by Congress in 2000 with the passage of Public Law 106-503. To the broad community with responsibility for reducing the effects of damaging earthquakes, this proposal represented a major step forward that would eventually lead to significant improvements in hazard assessment, structural engineering, loss estimation, and emergency management.

The past several decades have seen increased requirements from funding organizations for scientists to demonstrate not only that they are engaged in adding to the body of scientific knowledge, but also to demonstrate that scientific endeavors will ultimately provide tangible—and preferably quantifiable—economic benefits to the nation. It is in this context that the U.S. Geological Survey—the federal organization charged by Congress with monitoring natural hazards and providing natural hazard warnings—requested that the National Research Council (NRC) assess the economic benefits of improved seismic monitoring. The NRC committee was charged (Box ES.1) to give particular emphasis to the USGS plans for deployment of modern seismic monitoring instrumentation—the Advanced National Seismic System—predominantly in those urban areas that are most at risk from earthquake losses.

The committee heard presentations from a wide variety of experts and interested parties during its information-gathering meetings—from federal agencies, state agencies, local jurisdictions, private companies, and the academic community. During its deliberations, the committee grappled with understanding the extremely broad range of benefit types and with the variability in the degree to which the economic benefits of seismic monitoring can be proven and quantified. The committee recognized that there is a range of economic benefits that are extremely difficult or even impossible to quantify (e.g., assessing the value of reduced anxiety). Ultimately, the committee concluded that a compilation and description of

BOX ES.1
Statement of Task

An NRC ad hoc committee will provide advice regarding the economic benefits of improved seismic monitoring, with particular attention to the benefits that could derive from implementation of the Advanced National Seismic System (ANSS). In particular, the committee will:

- Review the nature of losses caused by earthquakes.
- Examine how improved information from seismic monitoring systems could reduce future losses in a cost-effective manner, taking into consideration the major impact-reduction approaches (for example, hazard assessment, building codes and practices, warning systems, rapid response, and insurance).
- Assess the capabilities for loss reduction provided by existing seismic monitoring networks, and identify how the ANSS and any other new monitoring systems would improve these capabilities.
- Describe concepts and methods for assessing avoided costs (both direct and indirect) that would result from improved seismic monitoring.
- To the extent possible, provide an estimate of the potential benefits that might be realized from full deployment of the ANSS.

the broad range of potential benefits—quantified where possible—would be the most appropriate and useful response to this aspect of its charge, recognizing that a base level of research and information is required before a rigorous and fully quantified estimate of potential benefits can be made.

The development of successful risk management strategies for earthquake hazards requires that the benefits from reduced uncertainty provided by improved seismic monitoring are integrated with the factors that influence risk perception and choice. The extent to which information from seismic monitoring networks can be used to reduce losses from future earthquakes depends, to a large degree, on providing this information to decision-makers and other end-users in an appropriate form, and on the extent to which these individuals and groups understand and make use of the information. With appropriate information, decision-makers—including public sector agencies, lenders, engineers, builders, insurers—can develop effective mitigation strategies to reduce the impacts of low-probability, high-loss earthquake events. Reducing the uncertainties in both the hazard and the risk through improved monitoring and research, and enabling policy-makers to gain an improved understanding of these uncertainties, also has considerable potential for improving the efficiency of emergency response planning and mitigation activities.

The potential benefits from improved seismic monitoring are quite varied. An important role of seismic information is to improve the accuracy (i.e., reduce the uncertainty) of building damage predictions and loss estimates, as the basis for more effective loss avoidance regulations, as well as enabling more effective emergency preparedness activities and improved earthquake forecasting capabilities. Each earthquake provides a unique opportunity to learn. Improved monitoring of future earthquakes will lead to a more complete understanding of geophysical processes, more effective hazard mitigation strategies, and improved emergency response and recovery.

As with all projects designed to reduce losses from natural disasters, the ANSS is expected to provide benefits in the form of avoided losses. Consequently, the costs of earthquake damage to the nation—without mitigation measures based on data and information provided by the ANSS—must form the benchmark against which the prospective benefits are assessed. Losses or costs associated with earthquakes fall into five major categories—direct physical damage (to buildings and infrastructure), induced physical damage (including fires, floods, hazardous material release, etc.), human impacts (death and injuries), costs of response and recovery (including first-responder and building inspection costs, etc.), and business interruption and other economic (social and environmental costs, etc.) losses. The most recent estimate of annual earthquake losses in the United States by the Federal Emergency Management Agency (FEMA) was $5.6 billion per year for buildings and building-related costs (after adjustment to 2003 dollars; building-related direct economic losses include repair and replacement costs for structural and nonstructural components, building content loss, business inventory loss, and direct business interruption losses), with a single significant earthquake potentially causing losses greater than $100 billion. Although concentrated on the West Coast, the risk of significant earthquake loss applies to many areas of the country. In recognition of the magnitude and extent of potential losses:

The United States should rank arresting the future growth of seismic risk and reducing the nation's current seismic risk as highly as other critical national programs that need persistent long-term attention, and it should make the necessary investment to achieve these goals.

Our understanding of the nature of earthquake hazards in the United States—the distribution, frequency, and severity of damaging ground shaking—is based on past damaging earthquakes as well as on the tens of thousands of small earthquakes that occur throughout the nation each year. Improved seismic monitoring networks will provide the basis for better characterization of this seismicity, so that the ground motion prediction models that underpin building codes and earthquake engineering design—the basis for safeguarding life and property—can more accurately reflect the complex nature of the hazard. In addition, any potential for the future prediction of damaging earthquakes will rely in part on seismic monitoring data.

Estimates of the extent of likely damage and the socioeconomic consequences from earthquakes are based on loss estimation models, which combine seismic hazard and vulnerability models with inventories of the built environment. Loss estimation models are contained in commercial software packages and in publicly available models, the most widely known and used in the latter category being the HAZUS model. HAZUS is a standardized, nationally applicable multihazard loss estimation methodology for estimating the impacts of disasters for the purposes of risk mitigation, emergency preparedness, and disaster recovery. All loss estimation models share a common structure—they are based on an estimate of the severity of the earthquake hazard, coupled with engineering estimates of the damage and loss to the infrastructure inventory in a particular region. In some loss model applications, the frequency of the hazard is also considered in order to provide the end-user with probabilistic loss estimates rather than scenario loss estimates. Output from the models typically includes the amount of expected damage to the built environment, economic costs of that damage (including business interruption costs), and estimates of injuries and deaths. Loss estimation models are used by insurers and reinsurers, government agencies, private businesses, the engineering community, and others. Improved seismic monitoring will enhance the accuracy of the data underpinning loss estimation models and reduce the uncertainty associated with model outputs.

The benefits from improved loss estimation model outputs include increased public knowledge, confidence, and understanding of seismic risk; better correlation between seismic risk and building code and land-use regulations; more efficient use of insurance to offset losses from disasters; and more accurate determination of the nature and growth of seismic risk in the nation. In addition, information about new and rehabilitated buildings and infrastructure, coupled with improved seismic hazard maps, will allow policy-makers to track incremental improvements in seismic safety through earthquake mitigation programs.

Improved earthquake hazard assessments combined with more accurate loss estimation models—both dependent on improved seismic monitoring—offer significant benefits for emergency response and recovery. These benefits include rapid and accurate identification of the event, its location and magnitude, the extent of strong ground shaking, and estimates of damage and population impacts. This information expedites hazard identification, promotes rapid mobilization at levels appropriate to the emergency, and facilitates the rapid identification of buildings that are safe for continued occupation and those that must be evacuated. These are tangible benefits to the emergency management community, and ultimately to residents of seismically active regions of the country. Although difficult to quantify, the ultimate benefits are lives saved, property spared, and human suffering reduced.

The integration of HAZUS loss estimation capabilities and USGS earthquake hazard information should be continued to track the growth of seismic risk in the United States, thereby further reducing the uncertainty associated with seismic risk.

The guidelines, standards, and codes available to earthquake engineers for the design of new structures and the rehabilitation of existing structures hold promise for protecting lives and the built environment against the largest expected earthquakes. However, perceptions that the up-front cost of mitigating the risk of earthquake damage is too high—combined with skepticism concerning the likelihood of earthquake occurrence, particularly in areas that have not experienced damage in historical time—leaves the country in a state of increasing seismic risk with the rapid expansion of the built environment. In order to make significant advances in arresting the growth of seismic risk, new analysis and design techniques are needed to better accommodate the expected ground motion. Current engineering design guidelines are mostly based on field observations that result in generalized and conservative procedures for controlling damage. The recent excellent performance of buildings where motion has been recorded provides a reasonable expectation that new techniques can be developed that will reduce the cost of seismic safety to more affordable levels. Seismic monitoring records hold the key to understanding how the built environment responds to significant earthquakes, and improved records offer the potential for fine-tuning the design process so that seismic safety requirements are adequately—but not excessively—met.

Determining the value of information has always been a challenge—it is not a tangible commodity and its benefits are often very subtle. Additional specific limitations apply to seismic monitoring information where the positive result of the information is avoided loss (e.g., in retrospective studies, it is difficult to isolate the contribution of seismic monitoring from other factors that influence the reduction in earthquake losses). Nevertheless, public policy decisions generally have to be made despite such limitations. The relative gains provided to society by improved monitoring information can be measured by the economic value of reduced decision uncertainty, assessed by comparing actions to be taken to manage the risks with and without improved monitoring.

It is possible, by using a series of assumptions, to determine a "ballpark" figure for earthquake losses that could be avoided by using improved seismic monitoring information as the basis for implementing improved performance-based earthquake engineering design. These assumptions relate to the value of the built environment within the United States, the cost of seismic rehabilitation and the number of existing buildings that need strengthening, and the annual expected loss from earthquakes compared with reduced losses when *higher* seismic design standards based on information from improved monitoring are applied. These calculations indicate a total loss avoided of more than $140 million per year, based on an estimate of reduced earthquake losses together with estimates of savings in construction costs that would accrue from the implementation of performance-based engineering design in those regions where improved seismic monitoring indicates that seismic design standards can be *reduced*.

Although it is possible to compile qualitative descriptions of the existing uncertainties and the potential economic benefits of improved seismic monitoring, there is insufficient existing research and information to provide a full quantitative assessment of such benefits. In effect, a certain level of seismic monitoring information—to be provided by the monitoring proposed for the ANSS—will be required before rigorous quantitative determination of the benefits can be made. The extent of the assumptions required to make the ballpark calculations for performance-based engineering design emphasizes the need for additional quantitative information before more precise estimates of the economic benefits of seismic monitoring can be determined.

After every damaging earthquake in the United States, data gathering and applied research should be sponsored—as a collaborative activity among the National Earthquake Hazards Reduction Program (NEHRP) agencies—to document how seismic monitoring information reduced uncertainty and provided economic benefits in both the long and the short term. Comprehensive reports should be published within one year after the event for short-term benefits, and within 10 years after the event for intermediate- and long-term benefits.

The relatively modest funding required to achieve a significant improvement in seismic monitoring capabilities should be considered in light of the potential for reducing the cost of constructing new facilities, strengthening existing structures to achieve proper performance, and avoiding losses from major damaging events. The approximately $200 million investment required for improved seismic monitoring infrastructure should be considered from the perspective of the more than $800 billion invested annually in building construction, the $17.5 trillion value of existing buildings in the United States, and the possibility of a $100 billion plus loss from a single, major earthquake in an heavily populated urban environment.

After assessing the considerable range of potential economic benefits from improved seismic monitoring that will be provided by full implementation of the ANSS, the committee concludes:

Full deployment of the ANSS offers the potential to substantially reduce earthquake losses and their consequences by providing critical information for land-use planning, building design, insurance, warnings, and emergency preparedness and response. In the committee's judgment, the potential benefits far exceed the costs—annualized buildings and building-related earthquake losses alone are estimated to be about $5.6 billion, whereas the annualized cost of the improved seismic monitoring is about $96 million, less than 2 percent of the estimated losses. It is reasonable to conclude that mitigation actions—based on improved information and the consequent reduction of uncertainty—would yield benefits amounting to several times the cost of improved seismic monitoring.

1

Introduction

Earthquakes are a continuing threat to the United States, capable of causing significant losses to the built environment and the economy that would equal or exceed those that occurred as a result of the 1994 Northridge earthquake, and human losses in excess of those experienced during the terrorist attacks of September 11, 2001. Recent estimates of the national earthquake risk project the average annual financial loss to be of the order of $5.6 billion[1] for buildings and building-related costs,[2] with losses from large earthquakes constituting a substantial portion of the expected losses (FEMA, 2001a). If utility and transportation system losses, business interruption and other economic losses, and the social costs of deaths and injuries are included, the number is almost certainly significantly higher. A single large earthquake could cause losses in excess of $100 billion to the built and human environment, more than twice the loss in the Northridge earthquake—the most costly U.S. earthquake to date (EERI, 2003).

The earthquake hazard is not evenly distributed throughout the nation. Although the greatest danger exists along the West Coast and in Alaska, 42 states have some degree of earthquake potential, and 18 are

---

[1]Unless otherwise noted, all figures are adjusted to 2003 dollars using the inflation calculator at http://data.bls.gov/cgi-bin/cpicalc.pl.

[2]Building-related direct economic losses (as estimated by HAZUS-99; see FEMA, 2001a) include repair and replacement costs for structural and nonstructural components, building content loss, business inventory loss, and direct business interruption losses.

considered to have areas of high or very high seismicity (USGS, 2005). Current estimates suggest that there are $17.5 trillion worth of structures in the United States (FEMA, 2004)[3]—the value of structures in high and very high seismic risk states is about $5.8 trillion (33 percent of the nation's total), and in all states prone to seismic damage the value is about $8.6 trillion (49 percent of the total). In population terms, 75 million people—including 46 million outside California—live in urban areas that have moderate to high earthquake risk.

Estimates of the extent of the seismic risk to the nation are based on three primary factors—the nation's inventory of structures (buildings, highways, pipelines, etc.), the potential damage extrapolated from performance in past damaging earthquakes, and the seismic hazard as determined from the geologic record and from instrument recordings of earthquakes that have occurred over the past century. This risk is growing steadily because buildings and infrastructure systems nationwide are being constructed without an adequate understanding of the seismic hazards that are present. The variation in seismic hazard across the nation is only now beginning to be understood as a consequence of the monitoring programs that have been in place for the last 50 years. Much more must be learned to quantify hazard levels properly, so that communities can understand their hazard and take appropriate, cost-effective steps to reduce their risk. Although efforts to quantify the vulnerability of the built environment are under way, current estimates have a significant amount of uncertainty because many recent earthquakes have not yielded the coupled seismic monitoring data and damage information that would enable improved vulnerability analysis.

Uncertainty—both epistemic and aleatory—exists concerning the frequency with which potentially damaging earthquakes occur in various regions of the country, the level of seismic hazard that results, and the damage that these hazards will do to the natural and built environment. The epistemic uncertainty, reflecting inadequacies in understanding the true state of nature, can be reduced by gathering more data through seismic monitoring. For example, improved seismic monitoring has the potential to improve current estimates of earthquake frequencies, of the median level of ground motion attenuation models (relating earthquake magnitude and distance to ground shaking levels), and of the median level of fragility models (relating ground shaking levels to building

[3]As estimated from HAZUS-MH (FEMA, 2004), based on 2000 census data, 2002 Dun & Bradstreet data, and 2002 Means replacement cost models. At the time this report was written, the valuation models in the HAZUS-MH software package were still being fine-tuned so future versions may produce slightly different exposure estimates.

damage). The aleatory uncertainty reflects the variability in phenomena that seems to be intrinsic and thus irreducible, although fundamental advances in science and engineering have the potential to identify causes of this variability and thus transform it to epistemic uncertainty. Improved seismic monitoring has the potential to provide more accurate estimates of aleatory variability—with implications for seismic design practice—because current building codes are based on a ground shaking level that has a 1/2,500 annual probability of occurrence, a level that is strongly influenced by aleatory variability in ground motion models. Improved seismic monitoring also has the potential to provide more accurate estimates of aleatory variability in earthquake frequencies and in fragility models.

As the nation's economy grows more interconnected and interdependent, moderate and larger earthquakes have the potential to cause significant national economic disruption and loss. A major earthquake in any of the vulnerable urban centers will have a ripple effect on the national economy and on the ability of American business to service and participate in the global economy. Recognizing that earthquake risk is a national problem and that improved seismic monitoring to mitigate this risk is a national responsibility, Congress created the National Earthquake Hazards Reduction Program (NEHRP) in 1977 to provide a coordinated national approach to addressing earthquake risk (Box 1.1). This program, which has been reauthorized by Congress eight times since 1977, is charged with furthering research on earthquake science, earthquake engineering, and social science research related to earthquakes, as well as implementation efforts related to improving building and infrastructure performance during earthquakes and more effective emergency response, recovery, and reconstruction (NRC, 2003c). Although NEHRP funding increased from slightly less than $70 million in 1978 to close to $100 million in 2002, this actually represents a declining trend in constant dollars.[4]

THE NATURE OF SEISMIC MONITORING

Much of what we know about the interior of the earth—and about earthquakes and their damaging effects—has been derived from seismic monitoring (NRC, 2003a). Earthquake monitoring is typically accomplished using both weak motion and strong motion seismometers, in association with geodetic networks that provide a measure of the deformation—or change in shape—of a region; this deformation is the result of strain caused by the same forces that give rise to earthquakes.

---

[4] The 2002 NEHRP budget was approximately half of the 1978 budget when expressed in constant 1978 dollars.

BOX 1.1
The National Earthquake Hazards Reduction Program

The National Earthquake Hazards Reduction Program (NEHRP) seeks to mitigate earthquake losses in the United States through both basic and directed research and implementation in the fields of earthquake science and engineering. Building on a foundation that was established following the 1964 Prince William Sound earthquake in Alaska and intensified after the 1971 San Fernando earthquake in California, the NEHRP was enacted by Congress in 1977 as the federal government's coordinated approach to addressing earthquake risks in the United States.

The program is managed as a collaborative effort among the Federal Emergency Management Agency (FEMA), the National Institute of Standards and Technology (NIST), the National Science Foundation (NSF), and the U.S. Geological Survey (USGS). Each agency's mission, although separate and distinct, has been integrated into a complementary program that emphasizes the transfer of research into practice and implementation.

FEMA, an agency within the Department of Homeland Security, works with states, local governments, and the public to develop tools for earthquake risk assessment and reduction and to improve policies and practices that reduce earthquake losses. FEMA had primary responsibility for overall planning and coordination of the NEHRP program from 1979 to 2004. In 2004, "lead agency" status was transferred to NIST as part of the 2004 reauthorization of the NEHRP program (P.L. 108-360).

NIST supports problem-focused research and development in earthquake engineering aimed at improving building codes and standards for both new and existing construction. NIST enables technology innovation in earthquake engineering by working with industry to remove technical barriers, evaluate advanced technologies, and develop measurement and prediction tools underpinning performance standards for buildings and lifelines.

NSF strives to advance fundamental knowledge in earthquake engineering, earth science processes, and societal preparedness and response to earthquakes. NSF programs such as the Network for Earthquake Engineering Simulation (NEES) provide a framework for collaborative and integrated experimentation, computation, and model-based simulation for the design and performance of civil and mechanical infrastructure. In addition, NSF supports basic research into the causes and dynamics of earthquakes, plate tectonics, and crustal deformation through programs such as EarthScope.

USGS monitors earthquakes, assesses seismic hazard for the nation, and researches the basic earth sciences processes controlling earthquake occurrence and effects. In addition to conducting engineering seismology studies of ground shaking, the USGS is also responsible for coordinating post-earthquake reconnaissance investigations.

Weak motion seismometers use very sensitive sensors to record the vibrations of seismic waves that travel through the earth. Weak motion systems typically are tuned to record a narrow frequency range of vibration, and these systems are ideal for recording small, local earthquakes (events located a few tens of kilometers from the recording station) or distant moderate to large earthquakes. **Strong motion seismometers** are low-sensitivity systems that can record the strong shaking that is actually or potentially damaging to man-made structures. Because the shaking experienced by a structure is a product both of the seismic source and the geologic materials on which the structure is built, these strong motion recordings provide the fundamental data used by earthquake engineers to design earthquake-resistant buildings.

Seismic monitoring encompasses the routine recording, analysis, and archiving of seismograms for a region. Typically, the product of such monitoring is a bulletin or catalog for each earthquake event containing source data such as location, depth, size (magnitude), and some indication of the type of faulting that caused the earthquake. Near real-time information—which aims to provide source location and magnitude information minutes to hours after an earthquake—is especially valuable as a tool for emergency managers. Such information can be used to improve initial estimates of the location and extent of damage. Later updates can provide more detailed information about the mainshock and any aftershocks, including improved location information and details such as the extent and directional attributes of the faulting.

Most existing seismic stations in the United States are analog, band-limited systems that by modern digital standards have significant technical limitations. These "narrow-band" systems were designed to provide a very narrow frequency band recording of ground shaking. The particular frequency band for each station was chosen based on the purpose of the network—high-frequency or short-period stations are typically tuned to record ground motions at approximately 1 Hz (or higher) and are used to record the timing of various seismic arrivals. However, this frequency tuning eliminates much of the spectra of the seismic signal, making it impossible to recover additional details of the earthquake process, such as the dimensions and magnitude of the earthquake and the orientation of the fault on which it occurred. Further, these systems have limited dynamic range, meaning that they cannot record the ground motions of large- or moderate-sized earthquakes that are located close to the recording station.

In contrast, modern digital seismic stations typically have broadband seismic sensors that are sensitive to a very broad range frequency band (typically 20 to 0.01 Hz) of ground motions, including the band from 20 to 0.2 Hz that has a significant effect on most structures. By including strong motion sensors together with sensitive seismometers, modern digital stations permit a very large range in size of vibrations to be recorded (i.e., the instruments have a large dynamic range), with the strong motion sensors staying on scale to provide useful data during even the strongest shaking.

EXISTING AND PROPOSED SEISMIC NETWORKS

Seismic monitoring activities in the United States today are based on a patchwork of regional seismic networks (see Figure 1.1) that use varying seismic instruments, often with limited fidelity and dynamic range. These networks are an evolutionary product of university-based networks that were installed in the early part of the twentieth century primarily to record the size and temporal-spatial distribution of major earthquakes. The Worldwide Standardized Seismograph Network (WWSSN) was established in the 1960s as a global network for monitoring nuclear explosions. During this same time period, regional seismic networks designed to monitor smaller earthquakes were established across the country. The principal factor driving the establishment of many of these regional networks was the need to characterize the potential seismic risk to critical facilities, especially nuclear power plants. During this time, the U.S. Nuclear Regulatory Commission (USNRC) funded the installation and operation of hundreds of seismic stations across the continent with the primary mission of monitoring the nation's seismicity (rather than recording strong ground motion). However, the USNRC abandoned these networks in the 1980s when it concluded that sufficient information had been gathered for plant design. The National Research Council (NRC) conducted a study in 1990 to assess the need for these networks and concluded that regional seismic networks were an essential component of the nation's strategy to decrease losses from earthquakes (NRC, 1990). That study proposed that the USGS should assume at least partial responsibility for a National Seismic System (NSS) to integrate data from the many regional networks. Unfortunately, funding for an upgrade of the NSS was extremely limited and most regional networks still continue to operate with equipment that is 20 or more years old.

At the heart of the nation's seismic monitoring capability is the U.S. National Seismic Network (USNSN), a "backbone" of seismic monitoring stations—operated by the U.S. Geological Survey (USGS)—that provides uniform coverage across the country and integrates data from its own stations and the more than 2,500 seismograph stations in regional networks. These regional networks provide information about earthquakes to the USGS National Earthquake Information Center (NEIC) in Colorado, which serves as a national point of contact for distributing earthquake information.

In 1997, the passage of Public Law 105-47 directed the USGS to assess the status of regional seismic monitoring networks in the United States, emphasizing the need for updated network infrastructure as well as expanded strong motion capabilities for urban area engineering purposes. In response, the USGS presented an assessment of the status, needs, and associated costs of seismic monitoring, proposing that an Advanced National Seismic System (ANSS) (USGS, 1999) would be required to meet the nation's needs for seismic monitoring (Box 1.2). This proposal included a comprehensive strategy to update and coordinate the nation's regional seismic networks to provide (1) public alerts within a few seconds of imminent strong earthquake shaking; (2) rapid assessments of the distribution and severity of earthquake shaking for use in emergency response; (3) warnings of possible tsunamis from offshore earthquakes; (4) warnings of volcanic eruptions; (5) information for correctly characterizing earthquake hazards and improving building codes; and (6) critically needed data about the response of buildings and structures during earthquakes for safe and cost-effective design, engineering, and construction practices in earthquake-prone regions (USGS, 1999). The distribution of ANSS seismic stations proposed for the nation's urban centers (Table 1.1) was based on relative seismic risk—a function of the population potentially exposed to strong ground shaking and the potential severity of that shaking.

Existing Weak Motion Monitoring Networks

There are approximately 1,700 stations[5] in the United States equipped with conventional weak motion seismic instruments (Figure 1.1) (CNSS, 1998). These stations are distributed among approximately two dozen regional seismic networks that are coordinated through the ANSS. Most of these networks are operated by universities or university-state partnerships, with partial funding from the USGS. Most of the stations (>1,500) are equipped with short-period seismometers, and the remainder with high-quality, broad band instruments. Most of these short-period instruments record only the vertical component of motion, and not the horizontal components which usually constitute the most damaging ground motions and form the basis for building codes. In addition, the instrument responses of most of the short-period instruments are poorly known. Accordingly, the short-period instruments are useful mainly for locating earthquakes and estimating their magnitudes and do not provide the ground motion recordings that are directly of use for earthquake engineering.

---

[5]Number as of September 1998.

BOX 1.2
Elements of the ANSS Proposal

The proposal for an Advanced National Seismic System described in USGS Circular 1188 (USGS, 1999) presents a plan for the modernization, expansion, and integration of the nation's earthquake monitoring networks. This proposal describes the capital investment (totaling $171.3 million in 1999 dollars; $189 million in 2003 dollars) and annual operating and maintenance costs required for this multicomponent physical and informational infrastructure (the following figures are adjusted to 2003 dollars):

- Expansion and modernization of the U.S. National Seismic Network from its present 56 stations to 100 modern seismographs distributed across the nation ($3 million)
- Replacement of 1,000 aging analog seismograph stations operating in regional seismic networks with modern digital instruments ($34.5 million)
- Installation of 3,000 ground-based, modern, digital strong motion seismographs in more densely populated areas at risk of strong ground shaking ($62 million)
- Installation of 3,000 modern, digital strong motion seismographs in buildings and structures (bridges, pipelines, etc.) ($62 million)
- Modernization of the National Earthquake Information Center (NEIC) and 20 regional network data centers ($24.3 million)
- Establishment of two portable arrays for aftershock monitoring and for special study deployment ($3 million)
- Annual operating and maintenance costs (including USGS overhead) for all elements of the fully deployed ANSS system ($51.75 million)

Small funding appropriations over several years have enabled the initial steps toward achieving the ANSS vision. The ANSS has been established by incorporating existing earthquake monitoring capabilities, involving collaborative arrangements with existing regional monitoring networks as well as installation of small numbers of modern digital seismographs and initial upgrades to NEIC capabilities. However, until significantly increased funding is available, the existing ANSS will be capable of achieving only a small fraction of the potential benefits described in this report.

As well as considering potential benefits that would result from implementation of the USGS vision for improved seismic monitoring proposed for the ANSS in both its present (embryonic) and proposed forms, the committee also considered the potential benefits that would result from an "ideal" level of seismic monitoring where critical and economically significant structures and lifelines in all urban regions of the country are monitored.

TABLE 1.1 Number and Distribution of Seismic Stations Proposed for the ANSS

Urban Area	Earthquake Hazard (%g[a])	Population (millions)	Relative Risk Factor[b]	No. of Urban Stations
Los Angeles, CA	88	15.4	5.1221	1,300
San Francisco, CA	99	6.5	2.4322	1,000
Seattle, WA	34	3.3	0.4241	600
Salt Lake City, UT	29	1.2	0.1315	400
Anchorage, AK	35	0.3	0.0397	300
San Diego, CA	25	2.6	0.2457	300
Portland, OR	19	2.0	0.1436	300
Reno, NV	33	0.3	0.0374	200
Memphis, TN	14	1.1	0.0582	200
St. Louis, MO	10	2.5	0.0945	200
Santa Barbara, CA	52	0.4	0.0786	100
Salinas, CA	43	0.4	0.0650	100
San Juan, PR	30	1.0	0.1134	150
Provo-Orem, UT	19	0.3	0.0215	100
Sacramento, CA	17	1.6	0.1028	100
Las Vegas, NV	12	1.1	0.0499	100
Chattanooga-Knoxville, TN	10	1.1	0.0416	100
Stockton-Lodi, CA	18	0.5	0.0340	60
Fresno, CA	12	0.8	0.0363	60
Charleston, SC	18	0.5	0.0340	60
Albuquerque, NM	11	0.7	0.0291	50
Eugene-Springfield, OR	14	0.3	0.0159	50
Evansville, IN	11	0.3	0.0125	40
Boise, ID	7	0.4	0.0106	50
New York, NY	6	18.1	0.4105	40
Boston, MA	5	5.8	0.1096	40
Total				6,000

[a]Hazard is expressed in terms of the severity of ground shaking—in percent of gravity—that has a 10% chance of being exceeded in the next 50 years.
[b]The relative risk factor is a function of the hazard factor and the population at risk; note that only a few stations are proposed for areas with very high populations but only low hazard.
SOURCE: USGS (1999).

Existing Strong Motion Monitoring Networks

There are approximately 1,400 strong motion recorders in the nation (Figure 1.2) (CNSS, 1998). Of these sites, approximately 200 are operated by regional seismic networks and the remainder are operated either by organizations that specialize in strong motion recordings (e.g., the California Strong Motion Instrumentation Program) or by private facilities (e.g., utility company monitoring of nuclear power plants, as required by the USNRC). The Consortium of Organizations for Strong-Motion

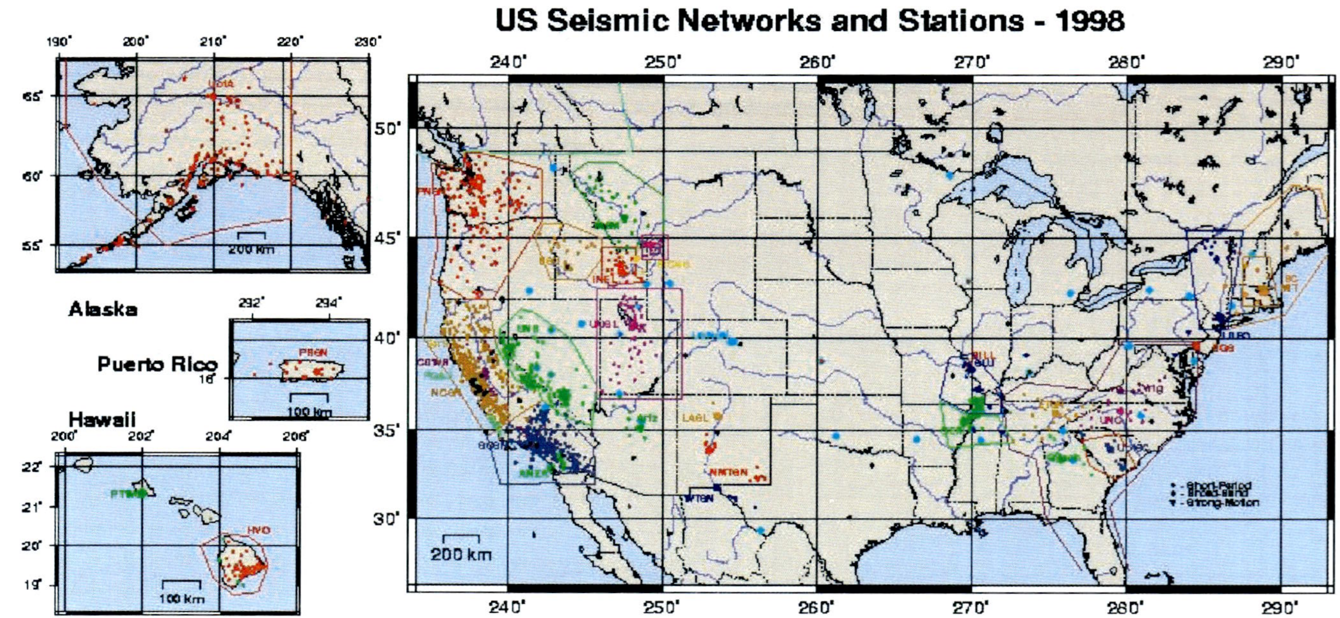

FIGURE 1.1 Distribution of seismic stations (dots) and seismic networks (marked by different colors) throughout the United States. SOURCE: Modified from CNSS (1998).

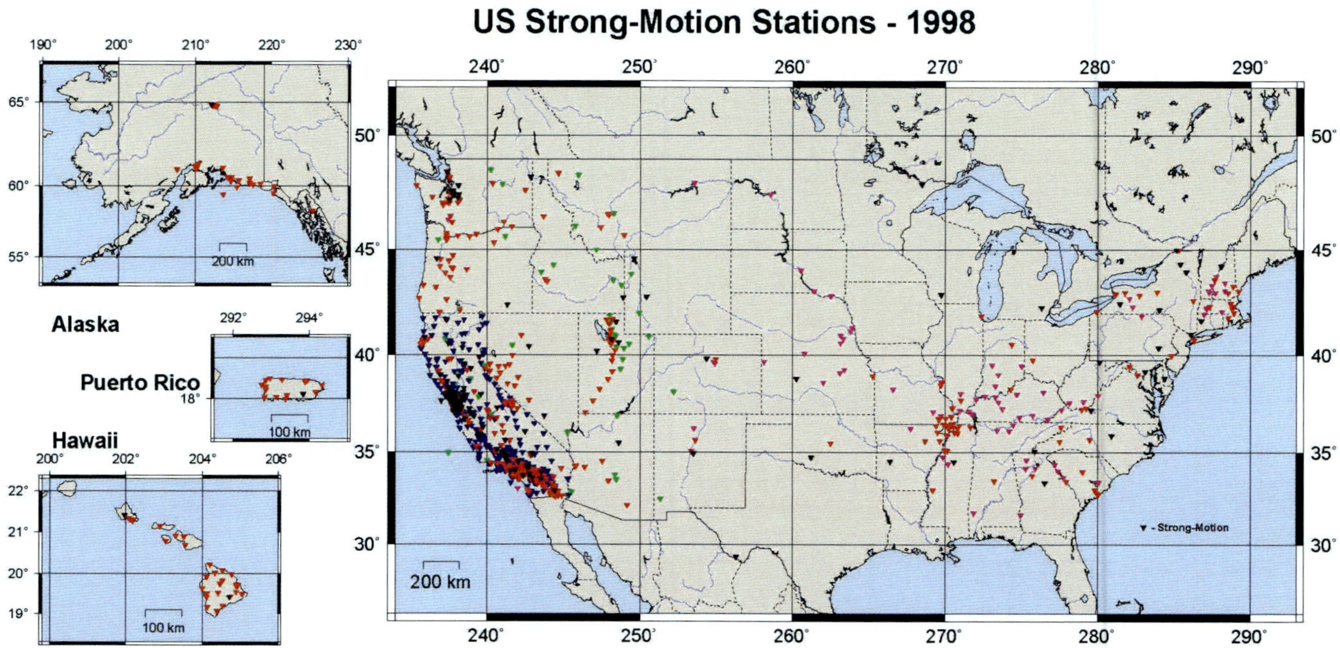

FIGURE 1.2 Distribution of strong motion seismic stations (dots) and seismic networks (marked by different colors) throughout the United States. SOURCE: Modified from CNSS (1998).

Observation Systems (COSMOS)[6] is an organization that represents most agencies and entities that are recording strong motion data and maintains a virtual data center that makes records easily accessible.

The deployment of additional strong motion seismic monitoring stations in urban regions has the potential to provide greatly improved descriptions of seismic hazard and risk. Even in the most densely instrumented urban regions, current strong motion recording stations at ground locations are too sparse to permit the compilation of reliable seismic zonation maps based on recorded earthquake data. Such maps have the potential to identify those parts of urban regions ("hot spots") that are especially prone to strong ground shaking during earthquakes, providing a basis for the prioritization of mitigation activities. Similarly, because of the small number of strong motion stations located in structures, few owners will have the benefit of a "proof test"—knowing the relationship between the level of ground shaking experienced by the structure and the level of ground motion for which it was designed. An ideal level of seismic monitoring—the ultimate extension of the present ANSS proposal—would permit the identification of ground shaking hot spots and provide for the proof-testing of all critical and economically significant structures and lifelines in all seismically active urban regions of the country.

EarthScope

In 2003, the National Science Foundation (NSF) provided initial funding for a major research equipment initiative called EarthScope (EFEC, 2003). This decade-long, research-oriented initiative seeks to understand the continental dynamics and structure of North America by integrating data from geology, seismology, geodesy, and remote-sensing observational facilities. EarthScope scientists have specifically identified the earthquake process as a primary scientific target, seeking to develop predictive models by unraveling the dynamic processes along faults (EarthScope, 2002). At present, EarthScope has three components:

1. A deep borehole observatory that is designed to measure physical conditions deep on a plate boundary. The San Andreas Fault Observatory at Depth (SAFOD) will be a 4-km deep hole drilled to intersect the San Andreas Fault (the plate boundary between North America and the Pacific).
2. A geodetic observatory designed to study the three-dimensional strain and deformation of the North American continent. The Plate Bound-

[6]See http://www.cosmos-eq.org.

ary Observatory (PBO) will consist of arrays of Global Positioning System (GPS) receivers, strainmeters, and seismometers.

3. A continental-scale seismic observatory (USArray) designed to provide a foundation for integrated studies of the continental lithosphere (the upper 150 km of the earth) and deeper mantle structure over a wide range of scales.

USArray is a research network that initially will provide results that complement the ANSS, and a component of this network will ultimately become part of the ANSS. USArray comprises:

- a *transportable array* of 400 portable three-component, broadband seismometers deployed on a uniform grid that will systematically cover the United States (with the exception of Hawaii);
- a *permanent array* of 40 high-quality, three-component seismic stations (the "Backbone Network") that will remain in place and become part of ANSS; and
- a *flexible array* of approximately 2,400 seismometers (a mix of broadband, short-period, and high-frequency sensors) that will be deployed within the footprint of the larger transportable array to record data for specific geologic targets.

USArray data will be used to image the details of the earth's structure beneath the nation. Stations will be deployed in many regions where there presently is only sparse coverage. However, the portable component of the USArray—which will provide the most dense station coverage—will be deployed at any one site only for approximately 18 months. Although there is likely to be serendipitous discovery in this short period of time, it will not fulfill the need for long-term monitoring. The permanent component of USArray—approximately 40 very high quality seismic stations—will become part of ANSS-USNSN and thus can be considered a leveraged contribution to the USGS system. The principal goal of ANSS is earthquake monitoring, an evolutionary and ongoing activity with a primary focus on continuously improved quantification of earthquake hazards. USArray, on the other hand, is a one-time comprehensive examination of the deep structure of the crust and upper mantle, primarily for research purposes. Thus ANSS is quite distinct from—yet complementary to—the USArray effort.

The Plate Boundary Observatory component of EarthScope is a geodetic observatory designed to study the three-dimensional strain field resulting from deformation across the active boundary zone between the Pacific and North American plates in the western United States. The PBO backbone will consist of 100 new and 20 existing GPS receivers that will provide a long-wavelength, long-period synoptic view of the entire plate boundary zone. An additional 16 GPS receivers will be located at USArray-supported stations in the USNSN. The backbone will cover western North America and Alaska at a receiver spacing of 200 km and eastern North America at a receiver spacing of 500 km. With the deployment of the PBO component of EarthScope, the USGS regional networks will take on renewed importance. A major focus of PBO's clusters of geodetic instruments will be the observation of transient deformation in the seismically and volcanically active areas of the western United States. In addition to surface observations of deformation, it is important to observe proxies for subsurface deformation—such as seismicity—through the use of regional networks. The areas covered by regional networks are essentially the same as those that will be covered by the PBO GPS-strainmeter deployments. In fact, this overlap is already being exploited, with GPS receivers planned to be co-located with regional network seismographic stations where feasible. In addition, PBO will be deploying high-bandwidth, three-component seismometers in each of the 175 borehole strainmeter sites. Consequently, PBO will be operating a 175-station regional borehole seismic monitoring network that will complement the existing regional networks operated by the USGS.

As both seismic monitoring programs—ANSS and the monitoring component of EarthScope—are implemented, there is an evolving relationship between the programs. One example is the location of PBO facilities within regional seismic networks to allow sharing of telemetry and maintenance costs. One of the ultimate goals of EarthScope is to understand the driving forces and consequences of plate tectonics to provide a better understanding of earthquake science—a goal shared with earthquake monitoring.

USES OF SEISMIC MONITORING

The transition path from data outputs produced by seismic monitoring instruments to information for decision-makers (e.g., emergency managers, earthquake engineers) requires analysis steps that differ depending on the type of seismic monitoring data and the area of decision. The different pathways are summarized in Figure 1.3 and described in more detail below.

Seismic monitoring networks provide the basis for hazard analysis and quantification, and the density of network coverage and instrument type determines how much will be learned from each damaging earthquake. Recent earthquakes have shown that existing seismic monitoring networks are too sparse to provide the essential information required to understand what happened during these earthquakes—the location and

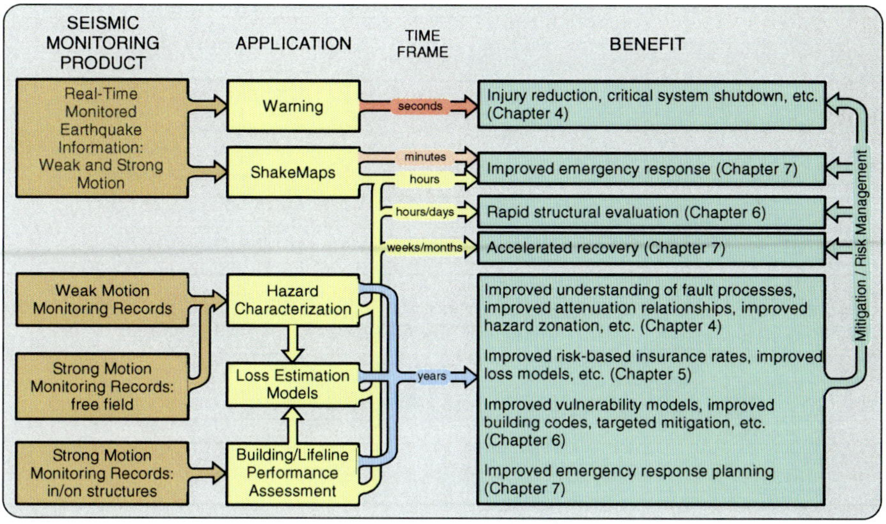

FIGURE 1.3 Flowchart summarizing the information path from seismic monitoring data outputs, through a range of applications, to ultimately provide the broad range of benefits described in this report.

nature of ground motion—and in particular what can be done to improve future construction and reduce vulnerability. This is true to some extent even in the three counties affected most by the 1994 Northridge earthquake,[7] where seismic monitoring was significantly upgraded following the earthquake. The potential economic benefits of improved seismic monitoring for engineering purposes are discussed in Chapter 6.

HAZUS—nationally applicable loss estimation software developed by the National Institute for Building Sciences (NIBS) for the Federal Emergency Management Agency (FEMA) (see Chapter 5)—combines a range of hazard information provided by the USGS (e.g., national maps of earthquake shaking hazard, real-time maps of ground shaking intensity following actual earthquakes) with engineering-based models of urban areas to estimate earthquake risk at both the national (FEMA, 2001a) and the regional or local scale.[8] A combination of improved earthquake loss estimation tools and improvements in data processing and communications

[7]Los Angeles, Orange, and Ventura Counties in southern California.

[8]See http://fema.gov/hazus/cs_main.shtml for case studies illustrating how HAZUS is being used to support hazard mitigation planning.

BOX 1.3
Emergency Response to the Chi-Chi Earthquake in Taiwan

At the time of the September 21, 1999, *Mw* 7.6 earthquake, Taiwan was in the midst of a seismic network upgrade project initiated in 1997 by the Central Weather Bureau (CWB). One component of this network is a Rapid Earthquake Information Release System based on 61 real-time digital accelerographs that are connected via telemetry to the CWB Seismology Center. This system, which was in place and operating at the time of the earthquake, provided notification of the magnitude, location, strong motion data, and an instrumental intensity map to 247 organizations nationwide within 2 minutes of the earthquake through paging, fax, and the Internet. These organizations included all fire and police agencies at the local, regional, and national government levels; ministries with emergency response and recovery functions; dam and nuclear power plant managers; scientists; and the news media. Receipt of information from the seismic network played a key role in early situation assessment at the national government level. Information received in real time by high-level officials in the Ministry of the Interior facilitated the mobilization of emergency response resources at all levels of government by 3:30 a.m., less than 2 hours after the earthquake occurred (Uzarski and Arnold, 2001). In contrast, it required 45 minutes after the 1994 Northridge earthquake to transmit the magnitude and location of the earthquake to emergency response agencies and nearly a week to generate a shaking intensity map. Although the seismic network in urban California can now provide earthquake information and mapping in real time, other parts of the nation—despite moderate to high seismic risk—do not currently have access to accurate and reliable real-time information products for a major earthquake.

during the last decade have enabled the creation and internet distribution of vital information within tens of minutes after an earthquake event. In the critical short-term period immediately after an earthquake, seismic monitoring data are used to create an initial "snapshot" of the emergency to initiate and prioritize appropriate response activities (as occurred following the 1999 *Mw* 7.6 Chi-Chi earthquake in Taiwan [Box 1.3][9] and

[9]*Mw* denotes the earthquake *moment magnitude* scale, which is related to the total amount of energy released in the earthquake. Improved seismic monitoring will enable more accurate and faster determinations of earthquake magnitude for significant events.

the 2001 *Mw* 6.8 Nisqually, Washington earthquake). Geographic Information System (GIS) based ShakeMaps, an internet-based real-time map information product generated by the USGS, show the location of strong ground shaking within minutes to tens of minutes after an event. The economic benefits for emergency response derived from the products of improved seismic monitoring, such as ShakeMaps, are discussed in more detail in Chapter 7.

Aftershock advisories in the days to weeks following significant earthquakes provide input to scheduling of operations that may be adversely affected by aftershocks, as well as real-time warnings to rescue crews during response activities. Tsunami watch and warning bulletins issued by the Pacific Tsunami Warning Center for submarine earthquakes and volcanic ash advisories for erupting volcanoes issued by the Alaska Volcano Observatory are additional examples of the benefits that seismic monitoring networks provide in the short term.

In the long term, data from seismograph networks and monitoring arrays in structures (buildings, roads, bridges, pipelines, utilities, etc.) are used to advance the fundamental scientific and engineering understanding of earthquake occurrence and effects. In late 1999, NSF launched the George E. Brown, Jr., Network for Earthquake Engineering Simulation (NEES) to accelerate the development of new seismic mitigation technologies. NEES is a large-scale, fully integrated, national resource that is in the process of shifting the emphasis of earthquake engineering research from its current reliance on physical testing to integrated experimentation, computation theory, databases, and model-based simulation. It involves 15 major earthquake engineering experimental research equipment installations networked through a high-performance Internet. Working in collaboration with the broader earthquake engineering community, NEES is expected to use advanced equipment and simulation capabilities to test and validate complex and comprehensive analytical and computer numerical models. The success of multi-institutional efforts such as the NEES initiative depends on the data recorded by high-quality networks. Strong motion recordings in the free field (away from structures) provide realistic input for engineering simulations and studies of structural response of buildings, bridges, and other facilities. Strong motion recordings made inside buildings, bridges, and other critical infrastructure provide information about structural response and damage. These data can be compared with the design parameters to evaluate possible damage states immediately after an earthquake and prioritize inspection and repair activities. In the longer term, these data are essential for developing new generations of earthquake-resistant building systems.

COSTS OF SEISMIC MONITORING

Any compilation of the economic benefits of improved seismic monitoring, as a component of a benefit-cost analysis (BCA; see Chapter 3), must also include an evaluation of the costs to the nation of improved seismic monitoring. These costs include the following:

- the costs of maintaining and operating the nation's existing seismic monitoring networks, including present federal contributions to the ANSS as well as nonfederal contributions to the plethora of state, university, and private networks now included within the ANSS;
- the costs of expanding the nation's monitoring, particularly with the deployment of modern digital seismographs as part of the full ANSS proposal, and the operational and maintenance costs after the instrumentation is in place; and
- a component of the seismic monitoring costs of EarthScope, representing those costs associated with the permanent "backbone" network of seismographic stations and the PBO regional borehole seismic network.

Financial support for the present system of seismic monitoring networks is complex. In FY 2003, the USGS provided $4.8 million to support regional networks. The network operators also receive support from state agencies and universities in the form of direct dollars and in services (e.g., technical support); consequently, it has been difficult to determine the total costs of operating the nation's existing seismic monitoring networks. A recent estimate by USGS of the operating costs for the Pacific Northwest Seismograph Network indicates that the USGS provided approximately 60 percent of operational funding for this network. In response to a request by the committee, the USGS compiled an estimate of total earthquake monitoring expenditure for FY 2004, concluding that the total of approximately $32 million ($31 million in 2003 dollars) was distributed among federal and state agencies as follows (Figure 1.4):

- USGS ($19.0 million)—encompassing the Earthquake Hazard Program ($14.6 million) and existing ANSS instrumentation ($4.4 million)
- State of California ($8.1 million)—encompassing the California Strong-Motion Instrumentation Program (CSMIP), California Integrated Seismic Network (CISN), and dam and aqueduct monitoring
- Other federal agencies ($2.3 million)—including the National Oceanic and Atmospheric Administration's (NOAA's) tsunami monitoring program and the Bureau of Reclamation, U.S. Army Corps of Engineers, Department of Energy (DOE), and General Services Administration monitoring programs for structure instrumentation

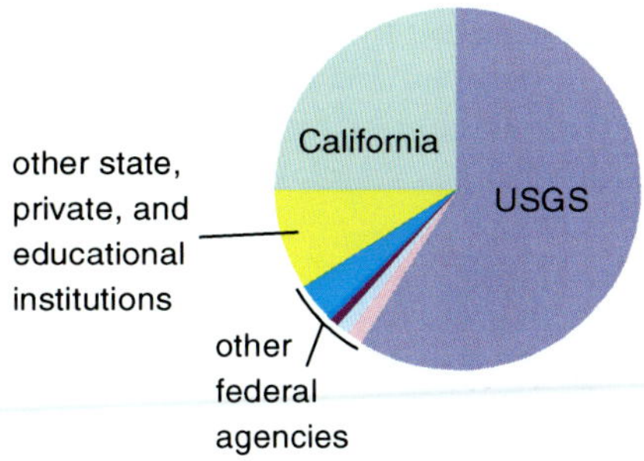

FIGURE 1.4 Distribution of FY 2004 federal spending for earthquake monitoring. SOURCE: Unpublished data provided by USGS (August 2004).

- A very approximate estimate of $2.9 million spent by other state and private industry monitoring programs, as well as educational institutions

These estimates do not include spending on the monitoring of volcanic seismicity (estimated to be less than $4 million); military and DOE monitoring related primarily to nuclear test verification; research monitoring (e.g., monitoring components of the EarthScope initiative); earthquake monitoring required for regulatory compliance (e.g., dams, nuclear power facilities); and "in-kind" contributions (e.g., the provision of facility space for instrumentation).

The costs to deploy, operate, and maintain ANSS instrumentation are presented in USGS (1999), amounting to $171.3 million ($189.2 million in 2003 dollars), with $47 million ($51.9 million) needed annually for operations. Congress authorized the ANSS in 2000, and the USGS now considers all of its existing seismic monitoring activities (except in situ volcano monitoring) to be components of the ANSS. Annual appropriations to begin implementation of the ANSS were $3.9 million in FY 2002 and FY 2003, increasing to $4.4 million in FY 2004 and $5.25 million in FY 2005.

Because of the short (18-month) deployment period for the research-oriented USArray transportable seismic system, it is not practical to consider the deployment, operational, and maintenance costs for these instruments as part of this analysis. However, the costs for the permanent USArray stations that will ultimately become part of the USNSN backbone—estimated to be $3.5 million for purchase and $1.1 million for operations and maintenance—are seismic monitoring costs that must be considered in a BCA. Purchase and installation costs for the high-bandwidth seismometers that will be emplaced at 175 borehole strainmeter sites as part of the PBO component of EarthScope are estimated at $1.4 million

(UNAVCO, 2005). The annual operation and maintenance expenses for this seismometer network have not yet been determined.

In summary, the cost to the nation of improved seismic monitoring has three components:

1. Annual costs for operating and maintaining existing seismic monitoring networks of approximately $31 million (recognizing that this is augmented by unquantified local support)
2. A total of $189 million for expansion and modernization of strong motion capabilities to establish the full ANSS, together with annual operating and maintenance costs of $52 million
3. A total of $4.9 million for hardware costs for the permanent USArray and PBO seismic monitoring components of EarthScope, together with annual operations and maintenance costs that have not yet been finalized

EXTENT OF LOSSES FROM EARTHQUAKES

Projects that are designed to reduce losses from natural or other disasters, such as improved seismic monitoring, are expected to provide benefits in the form of avoided losses or avoided costs. This means that the cost of such natural disasters—without mitigation measures such as improved seismic monitoring in place—must first be identified to establish a benchmark. This requires that for any particular area, the probability distribution of possible earthquake disasters and the consequent expected dollar losses must be calculated, necessitating a series of difficult estimates based on geologic and earthquake engineering projections. Although these are complex calculations, they must precede any complete estimation of project benefits. An assessment of the economic impact of improved seismic monitoring requires the identification of how the estimated benchmark losses would be reduced. This latter calculation would be undertaken for plausible increments of improved monitoring, if these can be identified.

Earthquake losses have been studied extensively by the NRC and others (e.g., see NRC, 1989, 1992, 2003b). In general, losses or costs associated with earthquakes fall into five major categories—direct physical damage, induced physical damage, human impacts, costs of response and recovery, and business interruption and other economic losses.

Direct Physical Damage

Direct physical damage is typically the largest contributor to overall losses and includes damage to buildings and infrastructure. In theory, this category of loss is the simplest to quantify, although in practice, detailed data are not often comprehensively compiled. Further, data may exist from a variety of sources. For example, the cost to repair a given structure may often be borne by more than one party (e.g., building owners, insurers). Nevertheless, the aggregate cost of direct physical damage has been estimated or measured in most recent earthquake events in the United States, as well as for some major historical events in other parts of the world (e.g., Tables 1.2 and 1.3).

Direct building damage may include structural damage, nonstructural damage, and damage to building contents and inventory. Other direct costs associated with loss of building function until it is repaired include relocation costs, lost rental income, lost wages, and lost income (NIBS/FEMA, 2002).

Infrastructure losses include damage to the transportation infrastructure (highways, roadways, airports, ports, light or heavy rail, buses, and ferries), as well as damage to utilities (electric power, water, wastewater, communications, oil and natural gas). In addition to the cost to repair such damage, the utilities may incur revenue losses associated with outages, costs associated with procuring alternate supplies, and in some

TABLE 1.2 Estimated Direct Losses in Significant U.S. Earthquakes

Earthquake	Fatalities	Economic Losses ($ million)	Insured Losses ($ million)
1906 San Francisco, CA	3,000	524 [10,700]	180 [3,680]
1989 Loma Prieta, CA	68	6,000 [8,903]	950 [1,410]
1994 Northridge, CA	33	44,000 [54,630]	15,300 [19,000]
2001 Nisqually, WA	1	2,000 [2,078]	305 [317]

NOTE. Original dollar figures are for "event year," with losses recalculated to 2003 dollars in square brackets.
SOURCE: Data from Munich Re Group (2000), except for Northridge fatality number from Peek-Asa et al. (1998) and Nisqually data from Munich Re Group (2002).

TABLE 1.3 Estimated Direct Losses in Significant Worldwide Earthquakes

Earthquake	Fatalities	Economic Losses ($ million)	Insured Losses ($ million)
1923 Kanto, Japan	142,800	2,800 [30,129]	590 [6,349]
1976 Tangshan, China	290,000	5,600 [18,107]	
1995 Kobe, Japan	6,348	>100,000 [>120,735]	3,000 [3,622]
1999 Izmit, Turkey	>17,000	>13,000 [>14,358]	1,000 [1,104]
1999 Chi-Chi Taiwan	2,400	>11,000 [>12,149]	>850 [>939]

SOURCE: Data from Munich Re Group (2000).

cases, regulatory fines. Indirect costs may also be associated with infrastructure damage, such as the economic cost of traffic delays or lost economic output due to closure in the face of utility outage.

In addition to potentially significant economic losses due to physical damage, rapid restoration of electric power systems is vital for both response and recovery. Direct costs to electric power utilities for the repair and replacement of earthquake-damaged equipment have been estimated to be ~$75 million for the 1989 Loma Prieta (*Mw* 6.9) and $183 million for the 1994 Northridge (*Mw* 6.7) earthquakes (Schiff, 1999). Because of the rapid restoration of electrical service and customer "resilience," such as the ability to make up lost production at a later date, these direct capital-related costs were much greater than the direct and indirect business interruption impacts to consumers (Rose and Lim, 2002; see Box 1.4).

BOX 1.4
The Costs of Electricity Disruption from Earthquakes

The results of two in-depth studies following the Northridge Earthquake shed some light on direct and indirect business interruption losses from electricity lifeline disruptions. Rose and Lim (2002) estimated the direct business interruption loss for the Los Angeles Department of Water and Power (LADWP) service territory as $109 million, including lost revenue to LADWP (although this excluded equipment damage to LADWP and its customers). This direct operating loss to the utility and its customers is roughly 50 percent of the estimated direct property damage to the LADWP system.

Most businesses are, however, highly resilient to short-term power disruption. Resilience factors include the ability to make up lost production at a later date (very prominent in non-service industries); time-of-day usage (the majority of the total outage of electricity took place before 9:00 a.m.); the fact that some aspects of production do not require electricity (most notably agriculture, construction, and transportation); electricity conservation; use of backup generators; and use of other types of energy or other inputs. Incorporating only the first three of these factors on a sector-by-sector basis into their model yielded a lower-bound estimate of only $5.6 million in direct business interruption losses, or about 0.4 percent of one day's production in Los Angeles County.

Tierney (1997a, 1997b) conducted a questionnaire survey and follow-up personal interviews of more than 1,000 businesses (responding) in the cities of Los Angeles and Santa Monica following the Northridge earthquake. Using Tierney's data on reasons for closing, duration of closure,

Larger earthquakes that impact larger areas, or occur closer to urban areas, will produce more extensive damage that can overwhelm the electric system redundancies and customer resilience that worked in the past for smaller earthquakes. As a result, unacceptable equipment losses, direct and indirect losses borne by customers, and lengthy disruption of service to the community are likely. The Applied Technology Council estimated that total indirect business losses from either a *Mw* 7.5 earthquake on the Hayward fault or an *Mw* 8.0 earthquake on the southern San Andreas Fault at Fort Tejon could be ten times greater than the direct losses experienced by utilities (ATC, 1991). Post-earthquake functioning of utility systems, particularly electric power service, is viewed by emergency responders and society in general as absolutely vital for rapid response and recovery activities following a major urban earthquake. Faster post-

and stated dollar losses sustained to businesses due to closure, Rose and Lim extrapolated the results to an estimate of $21.6 million of direct business interruption losses for Los Angeles County. Tierney's estimates are higher because she considered customers beyond the LADWP service area, her sample favored high-intensity impact areas and because Rose and Lim's simulation model may have overestimated resilience (most likely the ability to "recapture" lost production) in some sectors.

Rose and Lim also used an input-output model to determine indirect business interruption losses from the electricity outage. Noting that the outage was of very short duration and that inventory supplies were probably adequate to cover the lack of inputs of various goods and services that were not provided by the direct production decreases, the authors confined their estimate of indirect losses to "bottleneck" effects (see also Cochrane, 1997). The economy-wide multiplier in their analysis was 1.3 (meaning that indirect business interruption was 30 percent of direct business interruption), which resulted in an upper-bound estimate of total business interruption losses (now including the indirect component) of $142.1 million and a lower-bound estimate of $7.3 million.

Note that both the Rose and Lim and the Tierney estimates are much lower than estimates in a more recent study of electricity outages in the Los Angeles area. For example, the URS TriNet report (URS Group, 2001) estimated losses assuming a complete loss of electricity for a 24-hour period for the three counties of Los Angeles, Orange, and Ventura, and excluding nearly all resilience factors. Hence, it arrived at an estimate of $2.5 billion. The broad geographic coverage of the URS study would be relevant to only the most serious earthquake possible or if a system failure were triggered analogous to that of the Northeast electricity outage of the summer of 2003.

earthquake restoration of utilities enables a timely resumption of normal business operations. Disruptions of utility service and transportation networks due to earthquakes not only impact public safety and a community's ability to respond immediately after the event, but also affect the longer-term recovery. While the direct costs for the repair and replacement of earthquake-damaged lifeline facilities have been significant, there have also been very high indirect costs due to impaired infrastructure. One study of the 1994 Northridge earthquake included estimates of $6.5 billion of business interruption costs, of which $1.5 billion was ascribed to transportation network disruptions (Gordon et al., 1998). These costs were not captured in widely circulated structure replacement losses. The study also found that these losses could have been much higher were it not for the substantial redundancies in the road and highway network in the area.

A recent study on the effects of a large New Madrid earthquake suggests that the direct and indirect economic losses due to extended power disruption could be as high as $3 billion (Shinozuka et al., 1998). At the time of the study, very little empirical evidence was available to suggest that such a loss was even possible. However, in August 2003, a major power outage occurred over a large portion of the northeastern United States and Ontario, Canada. This event affected eight states, approximately 50 million people, and resulted in an estimated $4 billion to $10 billion of losses in the United States (ELCON, 2004). Although not earthquake related, the consequences are similar and demonstrate the effect that cascading failures can have on a regionally distributed power grid.

Induced Physical Damage

Earthquake shaking can cause damage to engineered structures and equipment, with the failure in turn resulting in a secondary or induced damage effect. Examples of induced physical damage include fire damage associated with post-earthquake fire or conflagration, potential flooding impacts associated with dam failure, hazardous materials release caused by building or equipment failure, and other environmental impacts. Induced physical damage impacts can be quite large, in some cases rivaling or even exceeding the cost of direct physical damage (e.g., fire losses in the 1906 San Francisco earthquake).

Human Impacts

The primary human impacts of earthquake consist of injury and death (together referred to as "casualties"), but also include displacement—often requiring long-term shelter—and quality-of-life issues such as mental health impacts or unemployment. Typically, casualty estimates are reported after most earthquakes, although the categories used and their definitions vary from event to event. Casualties resulting from the 1994 Northridge earthquake have been extensively studied (e.g., Seligson and Shoaf, 2003), and these are summarized in Figure 1.5. It is important to note that although the number of injuries and deaths is usually reported, the economic costs associated with these injuries and deaths often are not. Some of these data may exist in insurance files (e.g., health insurance, worker's compensation insurance), but historically the data have not been made publicly available. Recent estimates of economic costs associated with casualties in the Northridge earthquake have been as high as $2.2 billion (Porter et al., in press).

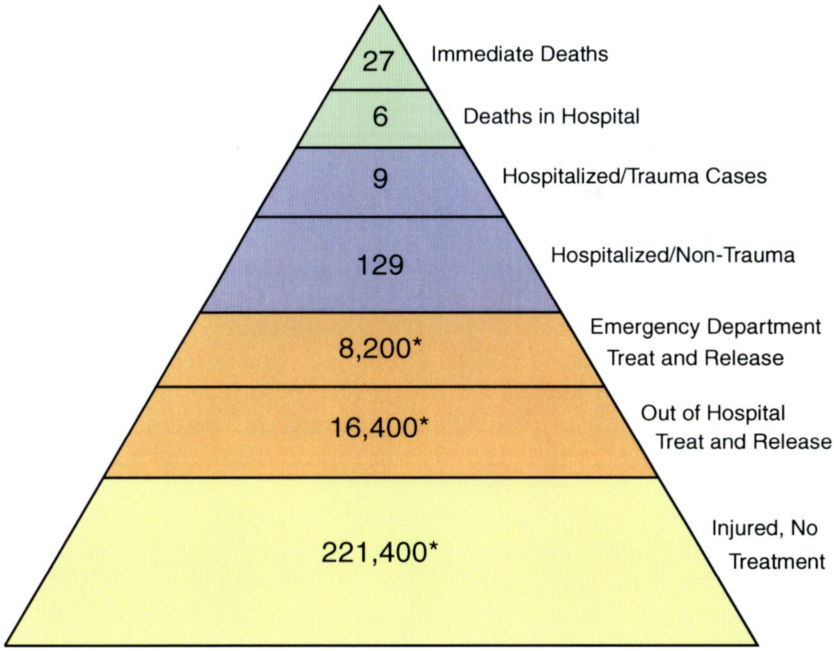

FIGURE 1.5 Injuries and deaths in the 1994 Northridge earthquake. Figures marked with asterisks indicate numbers of households; other numbers refer to individuals. SOURCE: Copyright 2003 from *Human Impacts of Earthquakes* by H.A. Seligson and K.I. Shoaf. Reproduced by permission of Routledge/Taylor & Francis Group, LLC.

Emergency Response and Recovery Costs

The cost of responding to an earthquake disaster can be significant, and in the United States such costs are often shared among local, state, and federal authorities. Emergency response costs include a wide variety of services:

- first-responder costs (personnel costs, including overtime), encompassing search and rescue, firefighting, and HAZMAT response; emergency medical services; police security at damage sites; and so forth; and
- service costs related to building damage, including post-earthquake building safety inspections (e.g., safety-tagging), emergency shoring and demolition, and debris removal.

A variety of recovery programs exist to facilitate individual, business, and community recovery. Such programs include loans (e.g., both privately funded and Small Business Administration loans), recoveries from private insurance policies, and grants (e.g., FEMA's Disaster Housing Assistance, Individual and Family Grants, and Hazard Mitigation Grants).

Business Interruption and Other Economic Losses

For many years, direct losses in the form of property damage comprised the measurement of economic consequences of natural disasters. However, during the past decade there has been a heightened awareness that other types of losses that may result from physical damage as well as other causes—such as direct and indirect business interruption, and environmental and social impacts—must also be considered (Mileti, 1999; NRC, 1999; Heinz Center, 2000; Ganderton, 2005). For example, direct business interruption can result from building or equipment damage, utility outage, lack of employees (due to injury, displacement, or transportation interruption), or supplier interruption. If enough businesses suffer interruptions, there can be a multiplier—or ripple—effect, indirectly impacting other economic sectors that may not have suffered direct damage of their own. This would include cancellation of orders by firms damaged or cut off from their transportation lifelines and the inability of other firms to sustain production because suppliers could not deliver critical inputs.

Indirect or secondary losses are those incurred in the days, weeks, or months following a disaster and include losses due to business interruption caused by infrastructure disruption (e.g., electric power, gas, water), reduction of critical services to residents in hazard-prone areas, and psychological trauma (Heinz Center, 2000). Indirect losses are difficult to document and are rarely measured after earthquakes, although methods exist to estimate potential impacts in actual and postulated events.

The 1994 Northridge earthquake is the best-documented earthquake event in U.S. history, providing an illustration of the magnitude of both direct and indirect costs. Table 1.4 provides a breakdown of available direct cost data for this earthquake, showing a total tabulated cost of $24 billion. In addition, data were not available for a number of significant costs—including the cost of insurance deductibles or uninsured losses—and these untabulated costs have been estimated at $20 billion.

TABLE 1.4 Direct Costs Associated with the 1994 Northridge Earthquake

Type of Cost	Total Estimated Cost ($ billion)
Buildings and Infrastructure	
Privately insured residential claims	$8.4[a]
Structures (coverage A) = $5.6 billion	
Appurtenant structures (coverage B) = $0.6 billion	
Contents (coverage C) = $2.0 billion	
Loss of Use (coverage D) = $0.2 billion	
Privately insured business claims (including a few public agencies that had insurance)	$4.1
Repair of transportation structures and roadways	$0.327
Utilities	$0.3
Public assistance[b]	$4.5
Emergency Response and Recovery	
American Red Cross	$0.036
Salvation Army	$0.001
Individual or family grant programs (including state supplemental grant and mental health)	$0.25
Hazard mitigation	$0.92
Small business administration	$4.03
Disaster housing and mortgage assistance	$1.2
California Employment Development Deptartment	$0.041
State Board of Control	$0.055
Subtotal	
Other costs (estimated) (insurance deductibles, uninsured losses)	**$20**
Estimated total cost	**$44**

[a]A more recent estimate of insured losses is $15.3 billion (Munich Re Group, 2000).
[b]Public assistance typically funds repairs to damaged publicly owned buildings and infrastructure.
SOURCE: Eguchi et al. (1998).

Consequently, the total direct cost of this earthquake was estimated to be as much as $44 billion (Eguchi et al., 1998). Indirect economic losses associated with transportation interruptions in this earthquake have been estimated to be an additional $1.5 billion (Gordon et al., 1998); no estimates have been published for other indirect costs.

Estimates of Future Earthquake Losses

Although recent worldwide earthquake losses exceed those experienced by the United States (compare Tables 1.2 and 1.3), it is reasonable to expect significant damaging earthquakes in the United States in the future—large earthquakes are inevitable in California, the Pacific Northwest, and other seismically active regions of the country. Further, population densities are increasing in our complex urban environments, and replacement of the nation's vulnerable building stock and infrastructure occurs very slowly. All of these factors contribute to an increasing risk of significant damage and loss in future earthquakes. FEMA (2001a) estimated expected annual buildings and building-related earthquake losses in the United States as $4.4 billion per year (see Figure 1.6), consisting of $3.5 billion per year in capital losses (building repair, lost contents and inventory) and $0.9 billion per year in income losses related to building damage (e.g., rental income losses, wage losses, business interruption). Using a Consumer Price Index (CPI) adjustment,[10] this expected annual loss total is $5.6 billion in 2003 dollars. In fact, EERI (2003) estimated that a single significant earthquake could result in losses greater than $100 billion. Although the bulk of annualized losses occur in California and on the West Coast, a relative earthquake risk assessment, as measured by the annualized earthquake loss ratio (losses relative to replacement value; see Figure 1.7), highlights other high-risk areas such as the central United States and Charleston, South Carolina.

Box 1.5 provides estimates as to how much earthquake losses from the 1994 Northridge earthquake could have been reduced had all of the buildings in the affected area been designed to the current seismic code. It also provides estimates of how a comprehensive seismic rehabilitation program could reduce economic and social losses from magnitude 7.0 scenario earthquakes on the Newport-Inglewood Fault in southern California and the Hayward Fault in northern California. These examples demonstrate the effectiveness of previous NEHRP work, including the use of information from seismic monitoring, in enhancing the effectiveness of building codes.

---

[10]See http://data.bls.gov/cgi-bin/cpicalc.pl.

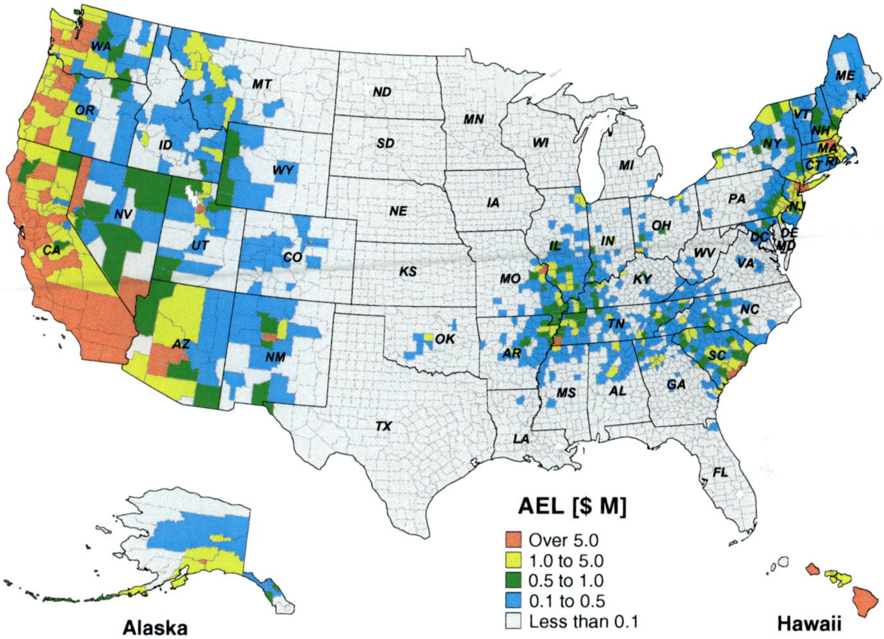

FIGURE 1.6 Map showing average annual buildings and building-related earthquake losses. Based on 1990 census data, this annual earthquake loss (AEL) was estimated to be $4.4 billion in 1994 dollars, equivalent to $5.6 billion in 2003 dollars. SOURCE: FEMA (2001a).

COMMITTEE CHARGE AND SCOPE OF STUDY

Over many years, a number of critical analyses by the NRC have recognized potential economic benefits from seismic monitoring information. Although it did not define the benefits of improved seismic information, a 1973 report contained an early and clear recognition that hazard reduction is beneficial to society (NRC, 1973). NRC (1980) contains an extensive discussion of the benefits and gains to society provided by seismic monitoring, and the broad categories discussed in this chapter are partially set forth therein. NRC (1990) continued the discussion by focusing on the benefits of a partnership between the proposed United States National Seismic Network (USNSN) and the existing independently operated networks.

The difficulty of assessing and quantifying the economic benefits of seismic monitoring—or indeed any monitoring programs seeking to mitigate natural hazards—has long been recognized by those charged with

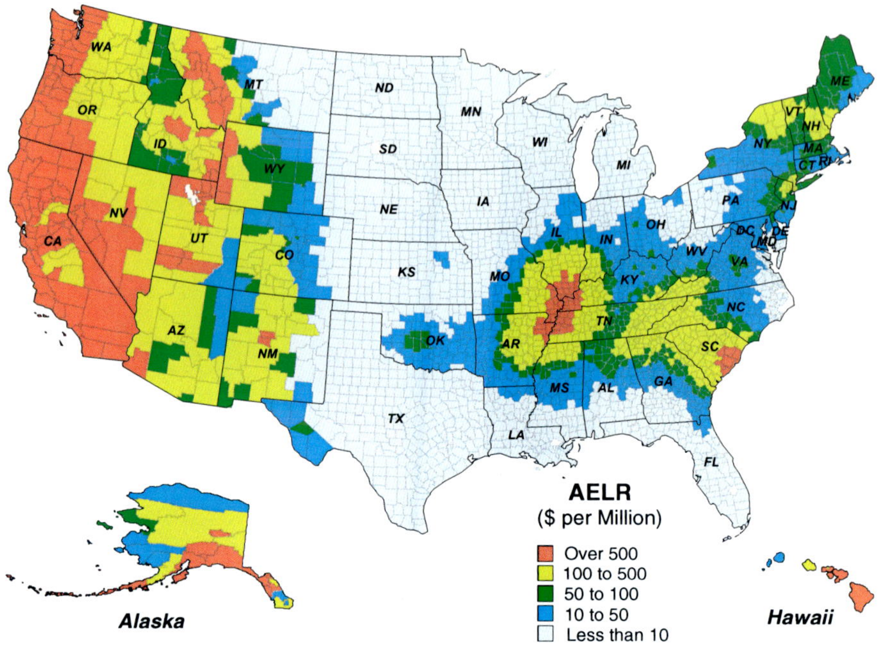

FIGURE 1.7 Map showing average annual earthquake loss ratio (AELR) at the county level. SOURCE: FEMA (2001a).

BOX 1.5
Building Codes and Earthquake Loss Reduction

Building codes have been recognized as one of the most effective tools for mitigating earthquake losses. The following examples illustrate the effectiveness of current building codes in reducing losses from earthquakes. Building codes, such as the Uniform Building Code and the International Building Code, became more effective as geoscience information from seismic monitoring and engineering information from damaging earthquakes were collected and analyzed as part of the NEHRP.

Urban areas contain a mixture of buildings, built at different times to codes with different earthquake requirements. FEMA (1997) estimated that the losses (building damage, contents damage, and income losses) in an event similar to the 1994 Northridge earthquake would have been reduced by 40 percent ($16.6 billion compared with $27.9 billion) if all buildings had been built to current high seismic design standards prior to the earth-

continued

BOX 1.5 Continued

quake. Had no seismic standards been in place, losses are estimated to be 60 percent greater than those for the baseline 1994 scenario ($45 billion versus $27.9 billion). In other words, the development and implementation of seismic design standards in building codes can lead to a reduction in earthquake damage by about a factor of 3. A 2001 FEMA report, based on the HAZUS-99 earthquake loss estimation methodology, examined the impact of seismic rehabilitation in reducing the economic and social losses from magnitude 7.0 earthquakes on the Newport-Inglewood Fault in southern California and the Hayward Fault in northern California (see Feinstein, 2001). Two scenarios were evaluated (all costs in 2003 dollars):

1. *Existing conditions or baseline*: representative of existing building stock in the area, a mixture of pre-code structures with no seismic rehabilitation and structures, either designed or rehabilitated, for various levels of seismic resistance
2. *Comprehensive rehabilitation*: the building stock is rehabilitated to the current seismic design levels for the region

		Baseline	Comprehensive Rehabilitation
Southern California	Building or contents damage	$69.5 billion	$51.3 billion
***Mw* 7.0**	Business interruption	$16 billion	$6.1 billion
Newport-	Displaced people	~400,000	93,000
Inglewood Fault	Shelter	~100,000	~29,000
San Francisco Bay	Building or contents damage	$32.6 billion	$22.2 billion
***Mw* 7.0**	Business interruption	$7 billion	$2.5 billion
Hayward Fault	Displaced people	~140,000	40,000
	Shelter	~30,000	~9,000

For these two examples, a *comprehensive rehabilitation program* could reduce building and contents damage losses more than 25 percent and business interruption losses by more than 60 percent. Major injuries and deaths, numbers of displaced people, and those requiring short-term shelter would also be reduced by more than 70 percent. For a more meaningful cost-benefit analysis of the mitigation measures in these examples to be undertaken, the cost of mitigation and building to these higher design standards, the value of lives and the cost of injuries, as well as the expenses associated with displaced persons, would all have to be specified.

justifying such activities. This has become increasingly important in the existing funding environment where scientific programs are required to demonstrate economic relevance. In response to the need for an independent statement describing the economic benefits of seismic monitoring, the USGS requested that the National Research Council conduct a review with the following charge (Box 1.6):

The review committee established by the NRC to address this charge received input from a variety of experts and interested parties during its information-gathering meetings—from federal agencies, state agencies, local jurisdictions, private companies, and the academic community. In this report, the committee specifically addresses the benefits provided by monitoring data derived from seismometers, but does not include the data provided by geodetic instruments and networks (e.g., GPS stations), which conceivably might be included within a broader definition of seismic monitoring. In addition, the analysis of seismic monitoring costs is focused on earthquake-related monitoring—strong motion instrumentation and that component of weak motion instrumentation related to distant earthquakes.

BOX 1.6
Statement of Task

An NRC ad hoc committee will provide advice regarding the economic benefits of improved seismic monitoring, with particular attention to the benefits that could derive from implementation of the Advanced National Seismic System (ANSS). In particular, the committee will:

- Review the nature of losses caused by earthquakes.
- Examine how improved information from seismic monitoring systems could reduce future losses in a cost-effective manner, taking into consideration the major impact-reduction approaches (for example, hazard assessment, building codes and practices, warning systems, rapid response, and insurance).
- Assess the capabilities for loss reduction provided by existing seismic monitoring networks, and identify how the ANSS and any other new monitoring systems would improve these capabilities.
- Describe concepts and methods for assessing avoided costs (both direct and indirect) that would result from improved seismic monitoring.
- To the extent possible, provide an estimate of the potential benefits that might be realized from full deployment of the ANSS.

One of the challenges facing the committee in its analysis of the potential economic benefits of improved seismic monitoring was to understand, first, how to quantify those benefits that are quantifiable and second, how to give appropriate credit to those benefits that cannot realistically be quantified but are nevertheless valid economic benefits to the nation. Given the time constraints and the nature of the project, the committee concluded that a compilation of the broad range of potential benefits—quantified where possible—was the most appropriate and useful contribution it could make. The committee's analysis and conclusions, described in the following chapters, discuss the contribution that scientific monitoring provides for decision-making (Chapter 2); the economic context for benefit calculation (Chapter 3); and the benefits of seismic monitoring information for hazard prediction and assessment (Chapter 4), loss estimation (Chapter 5), performance-based engineering (Chapter 6), and emergency response and recovery (Chapter 7). These various benefits are then integrated to form the basis for the committee's overall conclusions, presented in Chapter 8. Although the specific charge to the committee was to evaluate the economic benefits of seismic monitoring, it was clear to the committee that decisions regarding the allocation of resources for improved seismic monitoring must also take into account the costs of improved monitoring. Accordingly, some measures of the cost-effectiveness of seismic monitoring are presented in this report.

2

The Role of Seismic Monitoring in Decision-Making

This chapter describes the role of seismic monitoring in the decision-making process and provides examples of how seismic monitoring has been used successfully in the past. The three key components of decision-making that use seismic monitoring are:

Risk assessment—the role of monitoring in reducing risk and uncertainty.

Risk perception and choice—how individuals, groups, and organizations process information from seismic monitoring data and how this information influences their choices.

Risk management—the role of seismic monitoring as a contributor to strategies for dealing with earthquake hazards.

Risk assessment provides an understanding of the nature of risks—and their uncertainties—associated with disasters of different magnitudes, requiring input from the engineering and natural sciences disciplines. In the present context, this requires an understanding of the role that seismic monitoring plays in estimating earthquake risk, and how it can aid in reducing the uncertainties associated with these estimates. *Risk perception and choice* is concerned with the way earthquake monitoring data determines how individuals and organizations perceive their risk and make decisions in the context of the uncertainties surrounding the risk. *Risk management* describes the role of seismic monitoring in developing alternative strategies for reducing future losses and aiding the recovery process. An assessment of the contribution of seismic monitoring to disaster mitigation and management must be based on the integration of these three components. These elements—described in more detail in the following sections—provide the basis for evaluating the prospective benefits and projected costs of seismic monitoring in specific regions of the country.

RISK ASSESSMENT: THE ROLE OF MONITORING IN DEFINING RISK AND REDUCING UNCERTAINTY

Assessing the risk of earthquake damage to structures requires information concerning the

- type of structure and its response to strong ground motion and other seismic hazards,
- location of the structure in relation to earthquake faults,
- type of faulting, and
- overall distribution of strong ground shaking and its local modification by specific site geology.

Quantitative estimates of seismic risk are important for judging whether earthquakes represent a substantial threat at any location; they enable objective weighting of earthquake risk relative to other natural hazards and other priorities for making design and retrofit decisions (NRC, 1996). Earthquake risk assessment encompasses the range of studies required to estimate the likelihood and potential consequences of a specific set of earthquakes of different magnitudes and intensities. Scientists and engineers are asked to provide the key decision-makers—those who will use earthquake risk assessment data—with a description of the nature of the earthquake risk in specific regions as well as the degree of uncertainty surrounding such estimates.

The essential role of seismic information is to reduce the uncertainty in risk assessment over time and thereby increase its usefulness for emergency preparedness, loss avoidance regulation, private risk financing and insurance, and/or earthquake prediction. As improved monitoring provides increasing amounts of information, a more complete understanding of geophysical processes, more realistic models, and better-informed risk assessments will become possible. As the Advanced National Seismic System (ANSS) produces improved information, it will be possible to design better safety and regulatory programs, to generate improved ShakeMaps after earthquake events, and to improve earthquake prediction capabilities.

Within the range of geological and geophysical investigations conducted under the auspices of the National Earthquake Hazard Reduction Program (NEHRP), seismic monitoring plays a key role in the definition of the earthquake hazard—the foundation on which earthquake risk assessments are based. Seismic networks provide both parametric (e.g., earthquake origin times, locations, and magnitudes) and waveform or seismogram data. These data are used as the basis both for public safety decisions and for scientific and engineering research. Information about the locations of active faults and the size and frequency of damaging earthquakes allows decision-makers to specify appropriate design features of structures, using the seismic provisions in building codes.

One component of this work is the development of "earthquake design ground motion libraries," where strong motion records from a number of different earthquakes are processed in a consistent manner and made available to earthquake engineers and researchers in a web-accessible format for a variety of magnitude, distance, and fault types.[1] The collection of high-quality data at close distances is critical for validating and verifying the earthquake engineering models that are used in the construction of our urban environment. An additional component in the risk assessment process is the development of an understanding of building performance or capacity. Measurements of building response during actual earthquakes provide empirical information on seismic performance and can also provide information for evaluating the efficacy of current mitigation engineering practices. The goal of performance-based earthquake engineering is enhanced knowledge of how buildings respond to earthquakes so that structures can be designed to achieve specific performance objectives (above and beyond the life safety requirements described in current building codes) (see Chapter 6).

To illustrate how seismic monitoring can aid the risk assessment process, it is useful to consider a situation in which seismologists are asked to describe both the likelihood of earthquakes of various magnitudes occurring in the next 20 years (earthquake occurrence models) and the likelihood that the ground motions generated by these earthquakes will exceed some specified level (ground motion models). Seismologists can specify a probability distribution for a set of specific events, with bands of uncertainty that reflect the degree of confidence in these estimates (e.g., 95 percent confidence intervals). With increased seismic monitoring, scientists can refine both the earthquake occurrence models and the ground motion models, reducing the degree of uncertainty in those models. This information can then be used by engineers for estimating the likelihood of losses from earthquakes of different magnitudes as well as the degree of uncertainty surrounding these losses.

[1]See http://peer.berkeley.edu/smcat/.

An example of the usefulness of this kind of information is provided by the *Mw* 6.8 Nisqually, Washington, earthquake of February 28, 2001. The first deployment of ANSS strong motion instruments had fortuitously been made in the Puget Sound region only a few months before the earthquake. These strong motion recordings provided valuable information about site response, its correlation with surface geology, the effect of nonlinear soil behavior on site response, and the amplitudes of basin surface waves in Seattle and the surrounding region (Frankel et al., 2002). The deployment of portable aftershock recorders following the earthquake provided additional useful information (see Figure 2.1).

Immediately after a significant earthquake, there is a need to assess the extent and severity of damage and identify where emergency actions are needed. With the availability of ShakeMap, responders can pinpoint the areas of strongest shaking and focus their emergency response efforts quickly. One of the early uses of ShakeMap—a product developed in southern California in the years following the 1994 Northridge earthquake—was during the Nisqually earthquake. The ShakeMap software had just been installed in Seattle the previous month and was being used in a test mode, producing a ShakeMap within a few days of the earthquake. Subsequent aftershock monitoring emphasized the need for additional monitoring information. Comparison of ShakeMap contours of peak acceleration in Seattle using just the ANSS stations (left panel of Figure 2.1) compared with peak acceleration contours derived using the portable stations (described as "local network stations"; center panel of Figure 2.1) shows that the latter depicts a zone of strong shaking in central Seattle that was not identified using the ANSS stations alone. The latter panel has a much closer correspondence between the strong ground shaking in this region and the area that experienced damage, shown in the right panel, in which red dots indicate the locations of structural damage and ground deformation. By contributing to the seismic zonation of the Puget Sound region, especially in identifying locations that are potentially subject to large ground motion amplification or deamplification effects, seismic monitoring information enables seismologists and engineers to obtain more reliable estimates of the ground motion levels for which structures should be designed, thereby avoiding unnecessary conservatism in design.

One way to capture what is known and not known about a particular risk is to construct an "exceedance probability" (EP) curve, to specify the probabilities that certain levels of losses will be exceeded. Losses can be measured in terms of dollars of damage, fatalities, injuries, or some other unit of analysis. This can be illustrated with a specific example of an EP curve for an insurer with a portfolio of residential earthquake policies in a California city. Using probabilistic risk assessment, it is possible to combine the set of events that could produce a given dollar loss and then

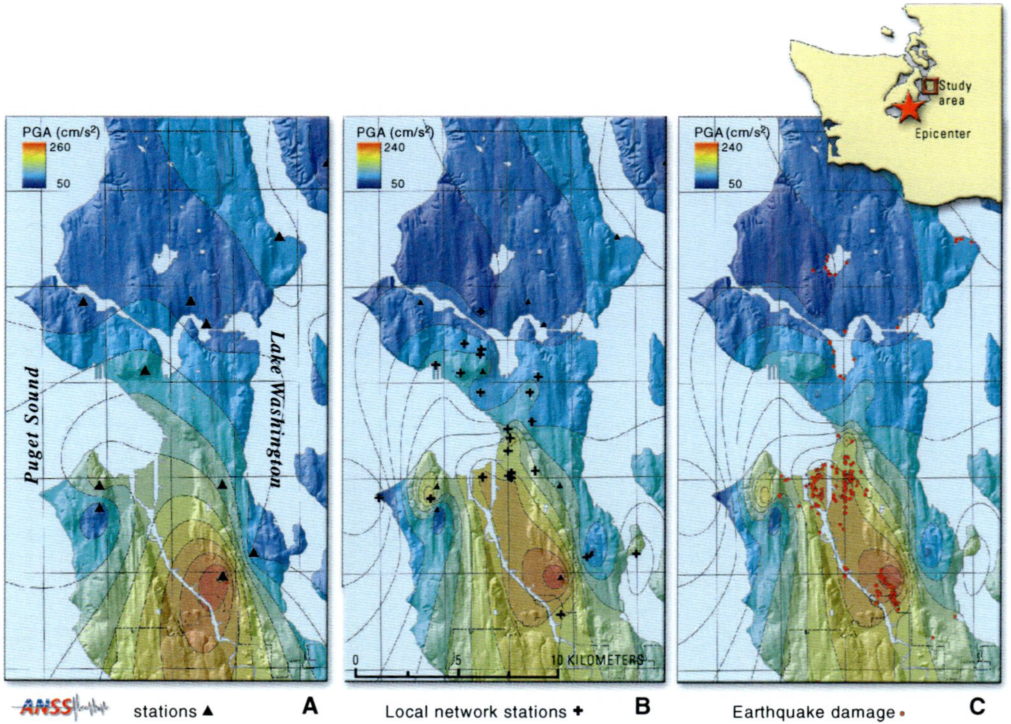

FIGURE 2.1 Comparison of ShakeMap contours of peak acceleration in Seattle from the 2001 Nisqually earthquake, using just the ANSS stations (left panel) and using the portable stations (center panel), shows that the latter map has a zone of strong shaking in central Seattle that was not identified using the ANSS stations alone.
SOURCE: USGS (2003b).

determine the resulting probabilities of exceeding losses of different magnitudes. The mean EP curve for such a situation is shown in Figure 2.2, illustrating that for the specific loss L_i, the likelihood that insured losses will exceed L_i is given by p_i.

Any interested parties can construct an EP curve—to depict the uncertainty associated with the probability of an event occurring and the magnitude of dollar losses (Figure 2.3)—to satisfy their needs and concerns. A company located in a hazard-prone area may wish to determine the likelihood that it will suffer direct dollar damage and indirect losses—such as business interruption—that exceed different magnitudes in order to determine how much insurance to purchase. A building owner may want to examine how specific protective measures will shift the EP curve downwards, to provide an indication of the impact such investments will have on future dollar losses to its structure.

As discussed further in Chapter 5, the large uncertainties associated with the likelihood and distribution of ground shaking, and with the damage to the built environment arising from shaking, significantly affect loss estimates. If these uncertainties can be reduced through seismic monitoring, they can lead to more cost-effective building design and construction decisions.

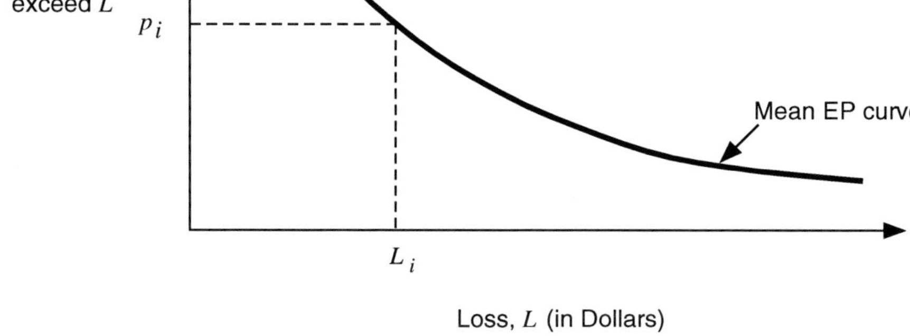

FIGURE 2.2 Sample mean exceedance probability curve, showing that for a specified event the probability of insured losses exceeding L_i is given by p_i.
SOURCE: Kunreuther et al. (2004).

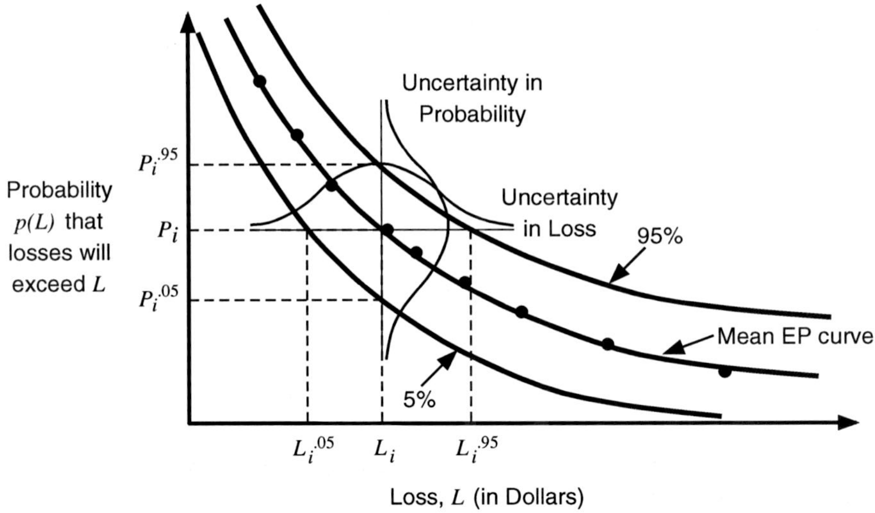

FIGURE 2.3 An example exceedance probability curve showing uncertainty expressed as 5% and 95% confidence levels. The curve depicting the uncertainty in the loss shows the range of values, $L_i^{.05}$ to $L_i^{.95}$, that losses can take for a given mean value, L_i ; this indicates that there is a 95% chance that the loss will be exceeded with probability p_i . Similarly, the curve describes the range of probabilities, $p_i^{.05}$ to $p_i^{.95}$, indicating that there is 95% certainty that losses will exceed L_i. SOURCE: Grossi and Kunreuther (2005).

In some areas of the United States, most notably in California, specific cost-effective risk mitigation measures are accepted (e.g., retrofit of unreinforced masonry buildings) and emergency response plans have well-articulated earthquake components. In these areas, scenario studies, which represent a horizontal slice through the EP curve in Figure 2.3, are used to estimate future losses, community vulnerabilities, and potential costs avoided through the implementation of mitigation strategies (see Box 1.5). Estimates of average annual loss, the area under the EP curve in Figure 2.3, are used by the NEHRP as a national and local measure and/or metric of seismic risk (depicted in Figure 1.6). However, in other parts of the country where damaging earthquakes occur less frequently, the risk is less well understood with the result that appropriate mitigation strategies are less clear and emergency response activities for earthquakes are less well established. Until the uncertainties surrounding the EP curve in Figure 2.3 are both reduced through continued monitoring and research and better understood by policy-makers, there will be either unnecessary or insufficient emergency response planning and inadequate mitigation of structures because the experts in these areas are unable to inform decision-makers of the probabilities and potential outcomes with an appropriate degree of confidence. This lack of confidence leads to a large gray area within which either over- or under-recognition of the extent of the hazard can seem reasonable.

RISK PERCEPTION AND CHOICE

Risk assessment focuses on the likelihood of certain events occurring, with damage and loss often able to be measured in monetary units. In the context of earthquake risk assessment, improved seismic monitoring has the potential to refine quantitative risk estimates. In contrast, risk perception is concerned with psychological and emotional factors—qualitative elements that have been shown to have an enormous impact on behavior. A set of pioneering psychological studies begun in the 1970s measured laypersons' concerns about different types of risks (Slovic, 2000) and showed that those hazards of which the person had little knowledge were perceived as being the most risky.

For a long time, the scientific community felt that it was appropriate to ignore the public's perception of risk if this differed significantly from its own estimates. There were many situations in which the public did not believe experts' figures because they were poorly communicated, because the assumptions on which they were based were poorly stated, and/or because there was little understanding of the reasons why experts disagreed with each other. For example, Expert 1 might say that there is "nothing to worry about regarding a particular risk," while at the same time the public would hear Expert 2 say that "this risk should be on your radar screen." The situation has changed in recent years, with an increased understanding of the importance of incorporating psychological and emotional factors in evaluating how the public assesses risk. Rather than basing choices simply on the likelihood and consequences of different events, as normative models of decision-making suggest, there is now recognition that individuals are also influenced in their choices by past experiences that may be unrelated to the actual risk associated with future events.

Surveys of homeowners in California support this point (Palm, 1998). These surveys suggest that the purchase of earthquake insurance is unrelated to any measure of seismic risk that is likely to be familiar to homeowners. Perceived risk, on the other hand, is a major predictor of earthquake insurance purchase. An illustration is provided by the Loma Prieta earthquake of 1989, which caused substantial damage to property in Santa Clara County and, to a lesser extent, in Contra Costa County. The percentage of earthquake-insured properties in Santa Clara County jumped more than 10 percent in the year following the earthquake. In Contra Costa County, the percentage insured increased from 22 percent in 1989 to 37 percent in 1993.

Even when no earthquakes have occurred in a given area, insurance purchase may increase if considerable concern is raised by media reporting. For example, there was a large increase in the demand for earthquake insurance in the New Madrid, Missouri, area when Iben Browning predicted that an earthquake would occur there in December of 1990 (see Spence et al., 1993, for description and analysis of this prediction). Even today, nearly 15 years later, a major insurer reports that more than half of its homeowners' insurance buyers in Memphis, Tennessee, also purchase earthquake insurance—despite the fact that there has not been any significant seismic activity in the area in recent years.

Individuals and businesses are not comfortable dealing with events in which there is considerable uncertainty regarding the likelihood of occurrence and the potential consequences. This aversion to ambiguity plays a role in the choices and decisions that individuals and businesses make with respect to high-impact, low-probability events such as earthquakes. Insurers who are trying to decide on premiums required for earthquake coverage provide an example—a series of empirical studies showed that actuaries and underwriters are so averse to ambiguity and risk that they tend to charge much higher premiums if the risk is poorly defined. Kunreuther et al. (1993) conducted a survey of 896 underwriters from 190 randomly chosen insurance companies to determine the premiums required to insure a factory against property damage from a severe earthquake. For the case in which both the probability and the losses were ambiguous, the premiums were between 1.43 and 1.77 times higher than if underwriters priced a nonambiguous risk. Similar results were observed in a study of actuaries in insurance companies (Hogarth and Kunreuther, 1989).

The problems associated with risk perception and choice are compounded by the difficulties that individuals have in interpreting low probabilities when making decisions. In fact, there is evidence that people may not even want data on the likelihood of a specific event occurring. A study of several hypothetical risky managerial decisions shows that when individuals are required to search for their own information, they rarely ask for any data on probabilities (Huber et al., 1997). One group was given a minimal description and the opportunity to ask questions. Only 22 percent of these respondents asked for probability information, and not one asked for precise probabilities. Another group of respondents was given precise probability information, and less than 20 percent of these respondents mentioned the word "probability" or "likelihood" in their verbal description of the factors impacting their decision-making processes.

To the extent that seismic monitoring increases the perceived likelihood of future earthquakes in an area and reduces the uncertainty and ambiguity surrounding these estimates, individuals residing there are more likely to pay attention to the potential damage from a disaster. Insurers are also likely to set their premiums closer to the expected loss because of the reduction in ambiguity provided by the improved forecasting.

Projects designed to reduce losses from natural or other disasters, such as improved seismic monitoring, are expected to provide benefits in the form of costs avoided. This means that the cost of such natural disasters—without mitigation measures such as improved building codes—must first be identified to establish a benchmark. This requires that, for any particular area, the probability distribution of possible earthquake disasters and the consequent expected dollar losses must be calculated, requiring a series of difficult estimates based on geologic and earthquake engineering projections. Each projected earthquake disaster event can be expected to cause structure damage and associated losses, business interruption losses, and infrastructure service losses. These three interact in complex ways, making the separate identification of each very difficult.

IMPACT OF MONITORING ON RISK MANAGEMENT STRATEGIES

In developing risk management strategies for earthquake hazards, the reduction in uncertainty associated with risk assessment due to seismic monitoring data must be integrated with the factors that have been shown to influence risk perception and choice. A framework for evaluating the impact of reductions of uncertainty has been proposed by Bernknopf et al. (1993) in the context of geologic map information that incorporates the type, structure, and engineering characteristics of a parcel of land. There is also now recognition of the need to define losses more broadly, to include both the direct impacts of a disaster (e.g., physical damage, direct business interruption, injuries and loss of lives) and the indirect losses (e.g., indirect business interruption, stress) (NRC, 1999; Heinz Center, 2000). This has made forecasting losses a more challenging task than when the focus was solely on direct property damage.

Improved Forecasting. To the extent that earthquake monitoring can lead to an improvement in the accuracy of forecasts, it has the potential to reduce losses from future disasters. Consider two homeowners in different parts of California (Regions A and B) who are considering investing in mitigation measures to reduce future damage to their homes. Suppose that in the absence of adequate seismic monitoring information, there is no distinction made between the two regions, both of which have an estimated probability p of a damaging earthquake occurring next year. Using this information as the basis for their choice, suppose that neither homeowner A nor homeowner B chooses to invest in mitigation. Both may be quite uncertain, however, whether this was the right decision to make. Now consider the case where the provision of seismic monitoring information enables differing likelihoods of damaging earthquakes in the two regions to be identified, with Region A having a probability $p_A > p$ and Region B having a probability $p_B < p$. Based on these more refined data, property owners in Region A are more likely to invest in mitigation measures and those in Region B may not be concerned with taking this action. Both groups would be more certain about their decisions and less worried about the consequences of any earthquake than they would be without seismic monitoring. Homeowners in Region A will have strengthened their residences and feel more secure physically. Homeowners in Region B now know that they are much less likely to have a damaging earthquake in the future than before monitoring was instituted, and they believe that mitigation is not a cost-effective strategy to follow.

Communicating Information on the Earthquake Risk. A number of studies indicate that people have difficulty assessing data regarding low-probability events (e.g., see Kunreuther et al., 2001). This poses challenges for effectively communicating information on these types of risk to the public. Improved seismic monitoring may lead to better communication of the risk because there will be less uncertainty regarding the likelihood of a future earthquake. It may also be possible to issue appropriate warnings about the dangers of earthquakes in particular regions of the country, leading to the adoption of risk reduction measures.

Using Economic Incentives. It is possible to use economic incentives to encourage individuals to take protective measures. Here again seismic monitoring plays an important role in decisions about whether to invest in mitigation. For example, greater certainty regarding the risk will lead insurers to price these policies closer to expected losses. At the same time, the premiums can more accurately reflect differences in risk between regions. Consider two identical homes in the above two-region example. The insurance premium for homeowners in Region A would now be higher than for those in Region B because $p_A > p_B$. Prior to the availability of adequate seismic monitoring information, the premiums would be the same in the two regions because they would both be based on p. Consequently, property owners in Region A should be able to get a larger premium reduction by investing in mitigation. The insurer now knows that if homeowners in Region A strengthened their houses, the expected earthquake claims payment would be less than it originally anticipated given the higher probability of a damaging earthquake in that region.

Incentive programs have been instituted in California to reduce losses from future earthquakes. Proposition 127, passed into law in November

1990, states that seismic retrofitted improvements to property completed between January 1991 and July 2000 will not be reassessed by the county tax assessor until ownership changes. The state—having concluded that these improvements constitute a significant reduction in the risks to life and safety—repealed the July 2000 cutoff, or sunset date (Chapter 504 of Statutes of 1999, introduced as AB 1291). To the extent that seismic monitoring can identify anticipated reductions of losses and lives saved in specific earthquake-prone regions, property tax reductions potentially can be designed so that they more accurately reflect the expected benefits of mitigation.

Building Codes. Building code regulations designed to mitigate seismic risk are desirable when property owners would otherwise not adopt cost-effective mitigation measures because they either misperceive their prospective benefits and/or underestimate the probability of a disaster occurring. When a building is substantially damaged or collapses, it may create losses to others in the form of economic dislocations and/or produce other social costs beyond the economic loss suffered by the owners. These losses would not be covered by the firm's insurance policy. A well-enforced building code helps reduce these risks and obviates the need for financial assistance to those who would otherwise suffer uninsured losses. By providing more accurate data on the likelihood of earthquakes through seismic monitoring, there can be a more systematic application of the seismic design provisions of building codes for different parts of the country.

DECISION-MAKERS/END-USERS AND THEIR ACTIONS

The decision-makers who will utilize the results of seismic monitoring in developing risk management strategies include builders and engineers, property owners, insurers and reinsurers, lenders, public sector agencies, and lifeline organizations, with the potential impacts of these risk management decisions affecting the lives of millions of people and trillions of dollars of the national economy.

Builders and Engineers. Developers, engineers, and contractors play an important role in the management of risk from earthquakes. Structures designed and built to high standards, combined with inspections by well-trained building officials, can provide good protection against casualties and property loss from earthquakes. Casualties and property loss are often attributable to inadequate design and construction practices. The problem of building and selling property in hazard-prone regions is exacerbated when uninformed design professionals and/or less reputable building contractors bypass costly seismic-resistant designs either that are not required by local codes or where the codes are not enforced (presented in more detail in Chapter 6).

Property Owners. Owners of commercial and residential structures that lack sufficient seismic resilience have a range of risk management strategies from which to choose. They can reduce their risk by demolishing a structure, retrofitting a structure to withstand earthquake loading, transferring part of their risk by purchasing some form of insurance, and/or keeping and financing their risk. Better seismic monitoring data will enable more informed decisions regarding the appropriate mitigation measures that should be adopted.

Commercial property owners' strategies to manage earthquake risks are different from those of residential owners. A commercial establishment must concern itself not only with life safety and insolvency issues, but also with the continued operation of its business activities following physical damage to its facilities and contents and/or infrastructure damage resulting in interruption of essential utility services (e.g., electricity, gas, water). Often there are extra expenses as a firm tries to remain viable after a catastrophe. Commercial establishments in hazard-prone regions are normally quite interested in purchasing business interruption insurance to protect themselves financially against these losses

Insurance Sector. An insurer provides protection against losses resulting from earthquake damage—from ground shaking and/or ground deformation—to those who opt to purchase separate earthquake coverage. Insurers also provide coverage for damage caused by fire following an earthquake to all who buy property insurance policies. Earthquake insurance can be purchased as added coverage to a homeowner's insurance policy; as a separate earthquake insurance policy; or, in California, through a state-run, privately funded earthquake insurance company—the California Earthquake Authority (CEA). In other states, earthquake insurance is provided solely by the private sector. Improved seismic monitoring will potentially provide better data to private insurers, reinsurers, and/or the CEA so they can more accurately price the coverage and manage their accumulations of risk. This will reduce the likelihood of the insurer's suffering unexpectedly severe financial losses following a major earthquake event and, in turn, should increase the availability and lower the cost of coverage.

Reinsurers accept and manage risk from insurers in the same way that insurers accept and manage risk from insurance buyers, and reinsurers must also price the coverage they offer and manage their accumulations of risk. They will also benefit from the availability of improved data, which will be reflected in increased reinsurance availability and lowered reinsurance cost.

Lenders. Lenders play a vital role in managing natural disaster risk. Except for the uncommon case in which the owner pays for property outright, banks and other financial institutions facilitate the purchase of a home or business by providing mortgages. The property is the collateral in the event that the owner defaults on the mortgage. Lenders thus have a vital stake in the risk management process, because they are unlikely to recover the full value of a loan on a property destroyed by catastrophe.

The 1994 Northridge earthquake, for example, generated $200 million to $400 million in mortgage-related losses in the Los Angeles area, and Freddie Mac[2] experienced an unprecedented number of earthquake-related defaults on condominiums (Shah and Rosenbaum, 1996). Seismic monitoring data have the potential to provide lenders with more accurate information on the risk. With these data available, banks and financial institutions have economic incentives to protect their investments by requiring risk-reducing measures and/or insurance as a condition for a mortgage.

Public Sector Agencies. Public sector agencies at the national and state levels should be able to design cost-effective earthquake mitigation and disaster preparedness programs that utilize the more accurate estimates of the risk obtained from seismic monitoring data. At the national level, the Federal Emergency Management Agency (FEMA) coordinates many of the planning and response activities related to catastrophes. FEMA has historically taken the lead in developing strategies for mitigation. For example, in December 1995, the agency introduced a National Mitigation Strategy with the objective of strengthening partnerships between all levels of government and the private sector to ensure safer communities. FEMA also provides funding to the Building Seismic Safety Council (BSSC) to develop the *NEHRP Recommended Provisions for Seismic Regulation of New Buildings and Guidelines for the Seismic Rehabilitation of Existing Buildings* (BSSC, 2004). These provisions and guidelines use the U.S. Geological Survey (USGS) national seismic hazard maps as the basis for defining the level of earthquake hazard for design and construction professionals. Improved seismic monitoring is the key to improving the accuracy of these maps, and public sector agencies would be able to use them in their design of seismic regulations and standards.

At the state level, an office of emergency services or a department of public safety promotes natural disaster preparedness. Additionally, seismic safety commissions have been established by earthquake-prone states to prioritize earthquake research and public policy needs. Building codes that include criteria for earthquake resistance and legislation for land-use management endeavor to reduce risk. At the local level, communities

---

[2]Freddie Mac is a government-sponsored entity that, along with Fannie Mae, provides a significant proportion of the funding for the secondary lending market by packaging mortgages and selling them to investors.

enforce building codes and have developed economic incentives, such as tax relief, for those who retrofit. Local communities develop programs to promote awareness, provide training, and encourage self-help actions through neighborhood emergency response teams. An example is the city of San Leandro, California, which has set priorities for retrofitting both unreinforced masonry buildings and older wood-frame homes. The city's *Home Earthquake Strengthening Program* is a comprehensive, residential, seismic strengthening program that provides homeowners with simple and cost-effective methods for strengthening their wood-frame houses to enhance earthquake survival. The program includes earthquake-strengthening workshops for residents, a list of available contractors, as well as a tool-lending library for homeowners should they wish to do the work themselves. Improved data from seismic monitoring will enable both states and communities to adopt building codes and design economic incentives that more accurately reflect the risk than those currently in place.

Lifeline Organizations. Lifeline organizations (including public and private utilities, transportation agencies, etc.) that provide water distribution and sewerage services, electric power, gas and liquid fuel, transportation, and communications play a vital role in the modern urban environment. Lifeline organizations are responsible for the resumption of critical services as soon as practical after an earthquake and have made significant investments to achieve this goal. The combined existing and planned expenditures for earthquake performance improvements for utilities and transportation systems in the San Francisco Bay area from 1987 to 2005 is estimated to be $15 billion.[3]

Improvements in the seismic design and post-earthquake operation of lifeline infrastructure are based on applied seismic research. The Pacific Earthquake Engineering Research Center (PEER) Lifelines Program (supported by the California Energy Commission, California Transportation Department [CalTrans], and Pacific Gas and Electric Company [PG&E]) and the Multidisciplinary Center for Earthquake Engineering Research (MCEER) (sponsored by the National Science Foundation and the Federal Highway Administration) are two examples of user-directed research programs that are actively involved in improving the safety and reliability of utility and transportation systems through the use of information collected by seismic monitoring. Networks of spatially distributed systems (transportation or utility systems) have a greater sensitivity to ground motions than individual (single-location) structures simply because the distributed system is affected over a larger area (especially for large-magnitude

---

[3]Presentation to committee meeting by L.S. Cluff, December 16, 2003, San Francisco, California: *The Role of Privately Owned Seismic Monitoring Networks.*

events). Reduced uncertainty will enable better allocation of resources for new design and construction or for mitigation of older facilities. Strong motion recordings used to develop time histories for seismic qualification testing of individual components or equipment (e.g., the IEEE-693-97 Standard)[4] provide critical information for improving electric utility safety and performance. Improved real-time monitoring of critical infrastructure (e.g., bridges, dams, rail lines, pipelines) will allow more efficient, prioritized inspections following earthquakes, as well as provide a means to monitor the long-term structural health of these facilities. In addition, federal regulations, such as the Department of Transport (DOT) Gas Transmission Pipeline Integrity Management in High Consequence Areas (49 CFR Part 192), require pipeline owners to collect information about earthquake faulting and ground acceleration as part of risk assessment and emergency response activities.

TECHNOLOGY TRANSFER

Technology transfer—outreach that explicitly seeks to apply new developments in science to solve practical community problems—is a vital component of the benefits from enhanced seismic monitoring. Technology transfer is distinct from public awareness or information campaigns, and the effective transfer or dissemination of new products and information to diverse types of users requires careful planning and well-articulated strategies.

The information and products derived from seismic network data are potentially useful to engineers, emergency managers, policy-makers, planners, insurers, the news media, and others. Whereas engineers use monitored seismic information to better understand damage caused to buildings and infrastructure by strong ground motion and to recommend retrofit options and mitigation strategies, emergency managers, planners, and insurers need to know the probability of damaging earthquakes in their communities for planning purposes. Emergency managers also need rapid information on magnitude, location, and ground shaking for effective response. The news media provide important public information during an emergency and expect to receive rapid, accurate, and reliable information from the seismic networks. To benefit from seismic network products at an optimal level, these products may require modification and adaptation to the specific needs of these various constituencies. A prerequisite for the effective use of network products by non-science users

[4]Institute of Electrical and Electronics Engineers, IEEE-693-97—Recommended Practices for the Seismic Design of Substations.

is an outreach effort designed to provide liaison between network operators and the users of network data. Interaction between data providers and data users should lead to an understanding of the available data—as well as their limitations—and should encourage opportunities for application of data products.

A simple but instructive example of such a process is one that took place between the operators of the California Integrated Seismic Network (CISN) and representatives of the major news organizations in California. Shortly after the development and introduction of ShakeMap in the version that currently appears on the Internet, outreach efforts were initiated to develop a wider audience for this seismic network product. Television news, because of its ability to reach large audiences with information on a widely felt or damaging earthquake, was considered an important target audience. Workshops were held in Los Angeles and the San Francisco Bay area that brought together network operators and news organizations to present the capabilities of ShakeMap and discuss how it might be adapted for use in television and print journalism. An important result from the workshop and follow-up activities was a Media ShakeMap that preserved the color-coded display of shaking, but eliminated measures of velocity and acceleration that were not likely to be understood by a non-scientific audience (see Figures 2.4 and 2.5). Improved seismic monitoring will permit the production of accurate, site-specific ShakeMaps throughout the United States that depict the post-earthquake conditions on the ground; present systems permit such accuracy only in southern California.

Technology transfer is also accomplished through a number of "intermediary" organizations that include—as part of their missions—the translation and dissemination of scientific research and new technology. Examples of such organizations are the Earthquake Engineering Research Institute (EERI); the Natural Hazards Research and Applications Information Center at the University of Colorado in Boulder; and the regional earth science and engineering centers funded by the National Science Foundation, the Federal Emergency Management Agency, and others. Technology transfer is accomplished through workshops involving scientists, engineers, and targeted users; by post-earthquake reconnaissance in which multidisciplinary teams investigate the impact of major earthquakes on the communities in which they occur; by the dissemination of publications that provide accessible information to users; and by formal training classes that disseminate new technologies for practical application.

PUBLIC INFORMATION BENEFITS FROM MONITORING

There are diverse demands for public information concerning earthquakes, ranging from curiosity (e.g., Was the shaking I felt because of an

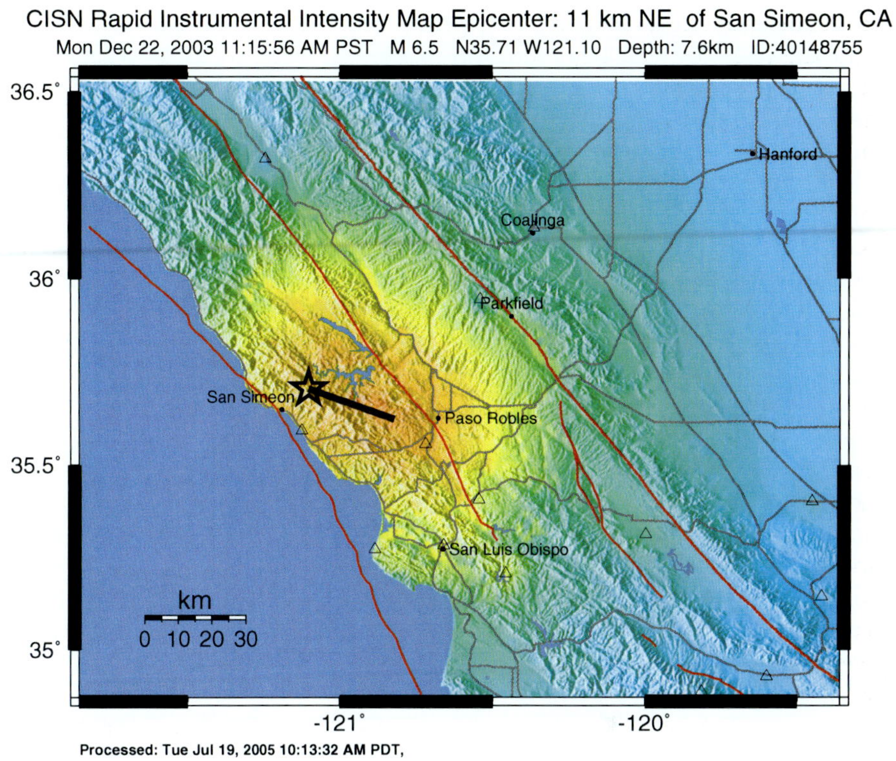

PERCEIVED SHAKING	Not felt	Weak	Light	Moderate	Strong	Very strong	Severe	Violent	Extreme
POTENTIAL DAMAGE	none	none	none	Very light	Light	Moderate	Moderate/Heavy	Heavy	Very Heavy
PEAK ACC.(%g)	<.17	.17-1.4	1.4-3.9	3.9-9.2	9.2-18	18-34	34-65	65-124	>124
PEAK VEL.(cm/s)	<0.1	0.1-1.1	1.1-3.4	3.4-8.1	8.1-16	16-31	31-60	60-116	>116
INSTRUMENTAL INTENSITY	I	II-III	IV	V	VI	VII	VIII	IX	X+

FIGURE 2.4 ShakeMap for the *Mw* 6.5 San Simeon earthquake of December 22, 2003.
SOURCE: USGS internet output . See http://earthquake.usgs.gov/shakemap/nc/shake/40148755/intensity.html.

earthquake?) to the very urgent need for emergency instructions associated with a warning of imminent danger (e.g., a tsunami alert) or after the occurrence of a damaging event (e.g., Where can I sleep tonight now that my house has been destroyed?). Seismic monitoring data provide the basis for communication of an increased seismic potential for a region—information that may be presented as a long-term statement of seismic risk or

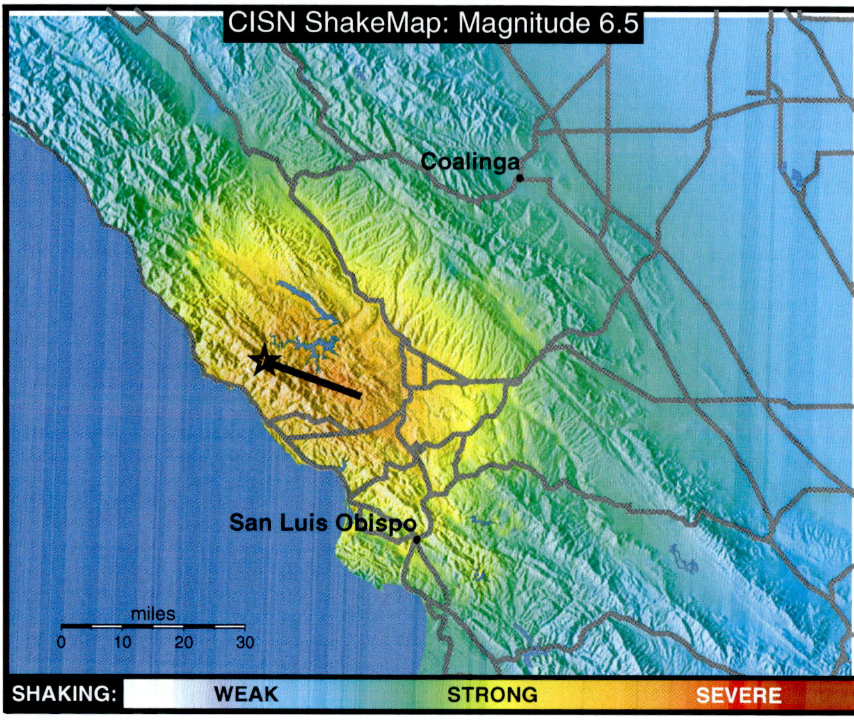

FIGURE 2.5 Media ShakeMap for the *Mw* 6.5 San Simeon earthquake of December 22, 2003; note the reduced amount of technical information compared to Figure 2.4.
SOURCE: USGS internet output. See http://earthquake.usgs.gov/shakemap/nc/shake/40148755/download/tvmap.jpg.

an announcement of an earthquake forecast in short or intermediate time frames. As tragically demonstrated by the 2004 Sumatran earthquake and tsunami, seismic monitoring information is of critical importance for communicating timely warnings that damaging tsunami waves may impact coastal communities.

More typically, public information about earthquakes takes the form of announcements of magnitude, location, damage, and possibility of aftershocks following an earthquake. With recent advances in technology, it is quite possible to obtain basic seismic information before the first public or news media inquiry is received. The availability of such real-time information, however, is limited to those areas with dense networks of modern digital seismic stations and instantaneous communication of data to a central processing site. These network hubs have become centers for the provision of public information following any earthquake that achieves newsworthiness—usually an event that ranges from being widely felt to one that impacts the entire community or region. Typically, after such an earthquake occurs, the news media converge at the network hub and scientists interpret what has happened for the assembled journalists.

The close relationship that has evolved in some regions between seismic network operators and the news media has facilitated the transfer of new knowledge and technologies to news organizations. As a result of these interactions, journalists become better-informed communicators of earthquake information to the public, and scientists become more aware of the needs of journalists for timely information in formats that are audience friendly. In addition, appreciation on the part of scientists of the critical link between the news media and the public has led to the development of specialized products designed to assist the media in communicating earthquake information to the public (e.g., Media ShakeMaps, real-time data feeds to news organizations).

It should be emphasized that readily available seismic information, the communication of this information to the public through the news media, and the close working relationship between scientists and journalists are not uniform in all seismically active regions of the country. In those areas that are poorly monitored, information about the occurrence of earthquakes is inevitably less timely and less accurate, and in these areas opportunities to improve the public's understanding of earthquakes, and the magnitude of the risk they pose, may be lost.

3

Conceptual Framework for Benefit Estimation and a Taxonomy of Benefits

The proposals for improved seismic monitoring put forward by the U.S. Geological Survey (USGS, 1999) are based on the premise—described in the previous chapter—that the provision of enhanced information will improve decision-making by reducing the uncertainties associated with risk assessment, with risk perception, and with choice and risk management. The relative gains provided to society by improved monitoring information can be measured by the economic value of reduced decision uncertainty, assessed by comparing actions to be taken to manage the risks with and without improved monitoring. Benefit-cost analysis (BCA) is one tool that can be used to evaluate alternative risk management programs. This chapter presents the conceptual basis for the specific categories of benefits—as input to a BCA—that are discussed in greater detail in later chapters.

On several occasions over the last few decades, the USGS has been asked to undertake benefit-cost analyses in support of program initiatives to generate improved earth science information. Typically, these requests have originated from the Office of Management and Budget (OMB). The ensuing studies were focused on estimating the economic value of information that is derived from earth science initiatives, using a decision-making framework that is supported by the theoretical principles of BCA and accepted estimation techniques. Similarly, this report also focuses on the issue of valuing seismic monitoring information within a benefit-cost framework. In this chapter, the following five questions are addressed:

1. What is the appropriate conceptual framework for valuing the prospective benefits of seismic information?
2. What is the nature of the information that is produced?
3. What are the various categories of benefits?
4. How can the benefits be measured?
5. Finally, has the necessary information been collected for a complete benefit-cost assessment of a seismic network?

An estimation of the societal benefits derived from seismic monitoring information has a number of components:

1. The framework for valuing benefits is based on the premise that enhanced information from a seismic monitoring network will lead to reduced uncertainty regarding the earthquake risk.
2. The scientific information gained from seismic monitoring will generate a variety of derivative benefits. As a seismic monitoring network captures ground motion for multiple events over time and space, gains in knowledge will assist with better decision-making in the future. Such gains can reflect increased accuracy as well as reductions in uncertainty.
3. The information produced by a seismic network is a "pure public good"—the information is available to all, and its use by one party does not detract from its use by others.
4. The major categories of benefits can be distinguished temporally as "immediate, near- and long-term, and cumulative gains in knowledge." In general, benefits may be derived from improved contributions to risk assessment, risk perception, and individual choice. An alternative perspective is that the benefits would include, but not be limited to, improved emergency response, enhancements in performance-based engineering, and increased potential for forecasting and predicting earthquakes.
5. A wide variety of hazard-prediction models and methods can be improved with enhanced seismic information, including ground motion and loss models.

The baseline for determining the economic benefits provided by improved seismic monitoring is the present situation in which the nation's seismic monitoring capabilities are distributed among a patchwork of essentially independent regional networks (described in Chapter 1), but with the important realization that existing funding levels are insufficient even to maintain present capabilities. Accordingly, any description of the economic benefits of improved seismic monitoring has to consider the incremental benefits in terms of

- the existing baseline situation, where existing funding levels will result in a gradual deterioration of the nation's monitoring capabilities;
- increased funding leading to maintenance of the existing networks and their integration into a revitalized U.S. National Seismic Network (USNSN);
- higher funding levels that permit the deployment of the Advanced National Seismic System (ANSS), as proposed in USGS (1999); and
- the ideal situation, in which the deployment of the ANSS is extended to provide instrumentation wherever it is needed.

Seismic monitoring data provide potentially derivable benefits that may be observed over multiple time frames. First, there are the immediate benefits after an earthquake (e.g., an informative ShakeMap can be used by emergency responders). This increment in more accurate information assists officials to deploy limited resources more rapidly and strategically to areas that have been identified as experiencing the greatest shaking (see Chapter 7). As a result, communities will experience more rapid—and consequently less expensive—restoration of services, resulting in reduced business interruption and cost savings.

Second, there are near- and long-term benefits from seismic monitoring information, related to the interval of time that allows society to react to the information in a strategic manner beyond the immediacy of an emergency response. The incremental benefits in this time frame principally reflect additional loss avoidance activities, beginning with property damage and running the course of all loss categories. Such loss avoidance would result either from information gained from a single event or from the accumulation of monitoring information over time. This accumulated knowledge can potentially result in an improved approach to the design and construction of infrastructure, the implementation of appropriate mitigation of existing structures, and/or the revision of building polices and regulations.

The third category of benefits is the accretion of knowledge. The accumulation of information from improved seismic monitoring potentially leads to a more complete understanding of the spatial and temporal physical processes associated with faulting and other sources of seismic activity. The accumulated record of weak and strong motion information could ultimately lead to some type of earthquake prediction capability (described in more detail in Chapter 4).

The limited time available for the committee to receive input, deliberate, and draw conclusions precluded the completion of a fully comprehensive BCA. Nevertheless, wherever possible, quantifiable benefits have been identified, evaluated, tallied, and compared to estimated project costs. Subsequent chapters demonstrate that there are numerous other economic benefits that will require a substantial compilation and analysis effort, and to assist this process the general principles of BCA that should be applied to improved seismic monitoring are outlined in the following sections.

BENEFIT ANALYSIS CONCEPTS AND APPLICATION

The economic efficiency of any project, including those owned and/or operated by the public sector, can be estimated by the application of benefit-cost analysis (BCA). Efficiency, as measured by a BCA, represents one dimension of any project's desirability as a component of an overall project evaluation. Efficiency is important in that it reflects the notion of resource scarcity. That is, are available resources being used in the most beneficial combinations from society's point of view, regardless of to whom the benefits or costs accrue? An elaborated BCA moves beyond a single efficiency criterion to address who pays and who benefits, thereby providing important information on the distribution of benefits and costs. While efficiency criteria are couched in dollar terms, when distributional aspects are added, the distributional measures are not necessarily comparable across projects. Thus, benefit-cost ratios (or differences) that describe alternative projects are scalars numerated in dollars that can be ranked, whereas distributional data are vectors of varying types of information and are not easily ranked within a project analysis or when comparing project evaluations.[1]

BCA is an elaborate accounting that observes basic economic principles. In this case, three basic ledger principles must be observed:

1. The ledger of costs and benefits must be specified with care, so that a comprehensive and exhaustive itemization is achieved.
2. Double-counting must be avoided.
3. Entries must not be mislabeled (e.g., the relatively common political suggestion that project benefits include the jobs created by a project when labor, in fact, is a cost).

To give just one (of many possible) examples of double-counting in BCA, consider the situation in which a municipality reduces its emergency

[1]Benefit-cost analysis focuses on an aggregate assessment (i.e., costs and benefits are simply tabulated with no consideration of to whom they accrue). However, from a broader public policy standpoint, the distribution of costs and benefits has always been important and its assessment has become increasing requested or required. This applies to issues such as whether benefits accrue only to a small segment of the population and whether those who receive benefits also pay for them. This has normative implications relating to fairness, as well as more pragmatic implications relating to public support for individual policies.

services for post-earthquake disaster assistance (e.g., by reducing the number of ambulances) and, as a result of this action, lowers its taxes. Any suggestion that both the reduced costs and the lowered taxes should be claimed as savings would constitute double-counting—both measure the same benefit and only one should be included.

All ledger items must be valued in dollar terms using market prices. Most benefits and costs are expected to occur in specific future years; at best, a distribution of each value for each future year can be made. The various possible realizations must each be weighted by their associated probabilities of occurrence. The resulting prospective benefits and estimated costs cannot be added for different years until properly discounted to a present value. Because costs change incrementally, it is important that plausible increments of any project be identified and that benefit-cost ratios that describe these increments are developed and reported. Where market prices do not exist because the commodity is not traded (e.g., clean air, seismic information), alternative methods—nonmarket valuation—must be used to estimate prices. The two broad categories of nonmarket valuation are "revealed preference" and "stated preference."[2] Of the revealed preference approaches, the "hedonic price method"[3] is widely used. It is recognized, for example, that residential property values respond to (capitalize) seismic information. By carefully controlling for other determinants of residential property values, it is possible to derive the dollar value of this information. The field of "contingent valuation" best represents the stated preference approach—contingent valuation studies use hypothetical markets to determine the willingness to pay for changes in risk levels. Because valuation is dependent on science information, any detailed contingent valuation analysis undertaken in the future must use the best available science information as the basis for understanding changes in risk.

CONCEPTUAL FRAMEWORK OF BENEFITS

As noted above, the basic ground motion and structure motion information that will be provided by improved seismic monitoring is a public good—that is, a product or service that can be shared by many users

[2]Revealed preference methods involve indirect valuation, based on observed market behavior reflecting choices (e.g., determining the value of a recreation site by using visitation, expenditure, and other data); stated preference methods involve direct valuation, which are undertaken by asking hypothetical questions—using alternative question formats—so that individuals assign value to changes in environmental services.

[3]The hedonic price method uses statistical attributes of market values to identify environmental or risk attribute components of overall property values.

simultaneously without detracting from its value to any one of them.[4] The nature of the good (or product) that emanates from improved seismic monitoring is essential for understanding why a public good framework is appropriate. Previous assessments of the societal value of geologic information—a similar type of informational good to the improved seismic monitoring goods—are applicable to improved seismic monitoring information; for example, Bernknopf et al. (1993) make the case that the information represented by geologic maps is in fact a public good (see Appendix A). The discussion contained in that report also applies more broadly to the information generated by improved seismic monitoring.

The important implication is that in identifying the various categories of incremental benefits in evaluating the economic benefits to be derived from improved seismic monitoring, summing the various benefit categories is appropriate because they are not mutually exclusive. As such, the appropriate evaluation of the economic benefits of improved seismic monitoring is its overall benefit based on the sum of its verifiable uses, rather than the benefits of any single independent use.

As noted in the preceding section, the theoretical basis for evaluating the benefits of geophysical information in general—and seismic information in particular—is efficiency gain. Improved seismic information provides the basis for better societal understanding and decision-making by reducing uncertainty. A number of studies have demonstrated the benefits and costs of improved seismic information in general (e.g., Bernknopf et al., 1990, 1997; Mileti et al., 1992; Olson and Olson, 2001). Although these studies have not specifically addressed improved seismic monitoring, they have included the benefits and costs of mitigation through building codes, microzonation information programs as enhancements to housing markets, and earthquake predictions.

The benefits that potentially are available as a consequence of improved seismic monitoring are quite varied. The fundamental role of seismic information is to reduce uncertainty over time and to increase the accuracy of emergency preparedness activities, loss avoidance regulations, and/or earthquake prediction. As increasing amounts of information are collected by improved seismic monitoring, the different vintages of seismic information (e.g., series of earthquake events—foreshocks, mainshocks, and aftershocks) will lead to a more complete understanding of geophysical processes, more realistic models, and better-informed risk assessments. As the improved seismic monitoring information evolves, it will be possible to generate improved ShakeMaps nationwide, to design

[4]The case of a pure public good, where, strictly speaking, it can be shared by an infinite number of users, is applicable to the USNSN/ANSS (e.g., see Randall, 1983).

better safety and regulatory programs, and to improve our earthquake prediction capabilities. The brief discussions of the four broad areas listed below are presented to set the stage for more detailed descriptions in later chapters.

Illustration 1—Improvements in Forecasting and Prediction. Additional seismic information will eventually improve the ability to microzone urban areas. Several types of incremental benefits can be envisioned. If the risk can be sufficiently differentiated geographically, there will be improvements in the efficiency of the operation of insurance markets. The benefit will be a reduction in the uncertainty of risk assessments and improvements in the accuracy of the information. Benefits will also accrue because of better forecasting and the potential for prediction. These types of benefit are discussed in more detail in Chapter 4.

Illustration 2—Improvements in Loss Estimation Models. An essential element of loss estimation models is their ability to adequately represent the frequency and severity of the earthquake hazard, as the basis for improving the efficiency of engineering design and providing better cost estimates for loss reduction. Because models of building vulnerability are based on limited performance data, improvements in the seismic hazard and building performance input data will reduce uncertainty in the estimation procedure. This type of benefit is discussed in more detail in Chapter 5.

Illustration 3—Improvements in Performance-Based Engineering. As a seismic network collects more information, new analysis will lead to better construction design criteria and lower construction costs. Because maps of the spatial distribution of seismic hazard can be improved, design criteria and structural damage mitigation will be better matched to local risk. This reduction in the uncertainty of how buildings will behave in an earthquake can be anticipated to reduce property damage and other losses. These types of benefits are discussed in more detail in Chapter 6.

Illustration 4—Improvements in Emergency Response Capabilities. If all bridges contained seismic monitoring capability, a determination could be made immediately following an earthquake as to which bridges were most likely to be damaged (based on fragility functions) and should be closed pending inspection and which were likely to be undamaged and immediately usable. Further, strong motion data recorded on bridges, together with damage data from these bridges, could be used over time to improve the accuracy of the fragility functions that relate ground motion to expected damage. This is an example of reduced uncertainty. The net benefit stems from the incremental savings in resources due to the fact that only damaged bridges, and not all bridges, would have to be inspected. Further, the transportation system could remain open or be opened sooner after a hazardous event, resulting in less business interruption. This type of benefit is discussed in more detail in Chapter 7.

TEMPORAL BENEFITS FRAMEWORK

The reductions in uncertainty that will translate into reductions in various types of earthquake damage—from the implementation of more appropriate improvements in the timing, location, and design of mitigation—can be categorized according to the timing of their benefits.

Immediate Benefits

The USGS (1999) carefully documented the immediate or time-critical qualitative benefits that would result from a revitalized USNSN-ANSS by demonstrating that real-time seismic monitoring systems offer the opportunity for society to take steps to reduce damage or loss of lives in advance of an earthquake. Even a few seconds of advanced warning before an earthquake may save lives by signaling people to "duck, cover, and hold" or to drive more slowly and avoid bridges or overpasses. Some machinery is more vulnerable to earthquakes while in operation, so real-time warnings could prevent damage by enabling the equipment to be turned off. Failure in one portion of a network can cascade into geographically widespread problems, as was demonstrated by the 2003 electricity blackout in the eastern United States. Chemicals or molten metals can harden, thereby spoiling the batch, damaging equipment if interrupted midprocess, and delaying the resumption of business. The avoidance of release of hazardous toxic materials from any medium is also a benefit. Table 3.1 presents examples of immediate benefits (the emergency response benefits are described in more detail in Chapter 7).

TABLE 3.1 Examples of Possible Actions That Yield Immediate Benefits

Action to Be Taken	Users	Incremental Benefit
Shut down critical lifeline systems	Utilities and their customers	Reduced damage to utility equipment and to customer equipment and materials
Shut down critical business equipment	Businesses	Reduced damages, avoidance of lost data
School children take cover	School officials	Reduced injuries
Avoidance of waste spills through real-time notification	Waste site managers and toxic waste mangers	Reduced damages
Protection of manufacturing processes	Manufacturers	Reduced losses in material batches

Near- and Long-Term Benefits—Avoided Loss Categories

As information accumulates from seismic networks, ground motion prediction models and loss estimation models will improve. The incremental benefits that result from these better models will lead to better engineering design parameters and more cost effective design, more detailed seismic zonation and land-use regulation, and so forth. As a result, there will be potential reductions in direct and indirect property damages compared to the situation without this improved information. There will also be reductions in direct and indirect business interruptions, reduced environmental damage, reduced iconic losses (both built and natural), and reduced human impacts. Furthermore, there is the potential to reduce infrastructure damage through improved seismic and structural "health" monitoring programs.

The first two columns of Table 3.2 provide a list of categories (adapted from Rose, 2004) and examples of incremental benefits from all potential uses of improved seismic information, although with an emphasis on near- and long-term applications. Direct property damage to buildings, contents, and infrastructure will normally be caused by ground shaking, although some property damage can result from subsequent deterioration caused by exposure to the elements. Indirect property damage is best exemplified by the impacts of fires, often resulting when ground shaking ruptures conveyances of flammable materials and is often further exacerbated by disruption of water delivery systems. Since property damage also typically sets in motion the other loss categories, mitigating property losses is fundamental to the near- and long-term incremental benefits of seismic monitoring.

One category of losses that can benefit from improved seismic monitoring is business interruption—diminished output of economic goods and services over some period of time from commercial enterprises caused by the earthquake. Business interruption can emanate from damage to physical capital but also from a cessation of other activity flows.[5] For example, a factory may be unscathed by an earthquake, but be forced to shut down if its electricity supply is cut off or its employees are unable to report to work due to earthquake-induced damage to transportation networks.

[5]The value of an asset is the discounted flow of net future returns from its operation. Hence, ordinarily the stock and flow measures represent the same thing, and at first pass, including both would seem to involve double-counting. The situation is, however, complicated for natural hazards. In this case, it is appropriate to count both stock and flow losses. The value of a destroyed piece of machinery is obviously a loss. Moreover, the inability to operate the business until the machinery is replaced represents a lost opportunity to the extent that it cannot be made up by overtime work in later periods.

TABLE 3.2 Time Frames and Applicability of Incremental Loss Reduction Benefits

		Potential Benefit Impact			
		Immediate		Near- and Long-Term	Knowledge-Based
Incremental Benefit Type	**Example**	**Warning**	**Emergency Response**	**Performance-Based Engineering**	**Forecasting**
1. Reduced direct property damage	Buildings, contents, pipelines	Limited	None	Extensive	Moderate
2. Reduced indirect property damage	Fire from pipeline damage	Moderate	Moderate	Extensive	Moderate
3. Reduced direct business interruption loss	Factory shutdown	Limited	Limited	Extensive	Moderate
4. Reduced indirect business interruption loss	Upstream and downstream ripple	Moderate	Limited	Moderate	Moderate
5. Reduced environmental damage	Toxic release to wetlands	Limited	Limited	Moderate	Moderate
6. Reduced other nonmarket damage	Public services, historic sites	Limited	Limited	Extensive	Moderate
7. Reduced social losses	Mortality, morbidity, sociological effects	Extensive	Extensive	Extensive	Moderate
8. Reduced government administrative cost	Ambulance service, fire protection	Limited	Extensive	Moderate	Moderate
9. Reduced emergency response cost	Disaster field offices	Extensive	Extensive	Moderate	Moderate

The time dimension also means that business interruption losses are highly dependent on private and public sector decisions and actions regarding recovery.

Additional losses stem from "multiplier" or "general equilibrium" impacts on chains of upstream suppliers and downstream customers of damaged businesses and those cut off from their utility lifelines or access by their employees or customers. These indirect effects can, in the case of large, highly interdependent, self-sufficient regional economies, be even larger than the direct flow losses (e.g., Webb et al., 2000).

In addition to more quickly identifying areas in which valves in pipelines carrying toxic materials might be shut off, there are several near- and long-term environmental benefits of improved seismic monitoring. Retrofitting wastewater treatment facilities and other sensitive structures will lead to decreased risk of drinking water contamination. Warnings that reduce the risk of fire will help avoid a deterioration of air quality. Performance-based engineering is also applicable to infrastructure other than buildings—the benefit of preventing physical damage to these systems is obvious and readily measured. However, interruption of the services that some of them provide is not so readily measured, since infrastructure services are typically not sold in the marketplace. For example, with the exception of toll roads, although the value of continued highway and bridge access is unpriced, this does not mean that it has no value (e.g., Gordon et al., 1998).

For many earthquake impacts that result from damage to structures and loss of businesses, there is a corresponding social impact that has economic implications (Heinz Center, 2000). Seismic monitoring information should result in benefits by reducing the impact on individuals, families, and communities as a consequence of reduced death and injury from improved building codes, improved dispatch of emergency services, and so on (Peacock et al., 1997; Enarson and Morrow, 1998). Broader social impacts that can be reduced include emotional stress and population dislocation, both of which have disproportionate impacts on marginal populations.

Prevention of the various types of losses already discussed can reduce government administrative costs, such as those associated with processing Small Business Administration loan applications and supervising emergency relief and recovery. Even if the government activity is a transfer, such as a loan, the administrative effort represents a real use of resources and is therefore a cost that improved seismic monitoring information can reduce.

The prominence of these various incremental benefit categories differs in terms of the uses of an advanced seismic monitoring system. The last four columns of Table 3.2 provide some "ballpark" estimates of the potential scope of four major applications of seismic monitoring—earthquake warning, emergency response, performance-based engineering, and earthquake forecasting. For example, its use for warning will reduce death and injury. However, warning will not reduce damage to structures, since they are immobile, but it can reduce damage to contents by facilitating emergency shutdown procedures. Monitoring systems can also reduce business interruption losses by giving electric utilities the opportunity to take precautions to avoid cascading outages. Moreover, loss reduction strategies can be fine-tuned to produce the mix of benefits considered to be in the best interest of society.

Knowledge Benefits

Increasing the number of earthquake strong motion recordings will contribute significantly not only to loss avoidance as discussed above, but also to improving society's capabilities—through advances in underlying knowledge—in performance-based engineering, seismic zonation programs, and earthquake prediction. The accumulation of knowledge is typically viewed as an ongoing and long-term process, and it is this accumulation of knowledge that makes improved loss avoidance and emergency response and recovery gains possible.

Two examples illustrate situations in which a more complete set of ground motion records will enhance knowledge benefits. First, current USGS probabilistic ground motion predictions near major active faults are adjusted before they are used in the National Earthquake Hazards Reduction Program (NEHRP) design ground motion maps, so that they depict ground motion levels that are more compatible with observed damage in earthquakes. The ground motion models used in USGS maps are based on the few recordings at close distances to large earthquakes and are thus highly uncertain predictions. Effectively, the lack of a more substantial set of near-fault strong motion recordings makes it difficult to reliably implement performance-based engineering, since the levels of damage to structures observed in the near-fault setting seem to contradict the high ground motion estimates currently used. Second, seismic zonation will be improved with more complete ground motion records. With an improved understanding of the underlying geological structure, narrow regions of higher vulnerability can be identified within broader regions that are now considered to have equal exposure to the hazard. Improved monitoring will provide improved definitions of seismic "hot spots" for design and land-use plans, thereby contributing to more rational (and less risky) urban development.

BENEFIT ESTIMATION PRINCIPLES AND PROCESS

An evaluation of benefits resulting from the proposed improvements to the nation's seismic monitoring networks must be based on the prospective benefits from the additional information that will be forthcoming. Estimating the value of information has always been a challenge because it is not a tangible commodity and because its benefits are often very subtle. This is compounded by both hazard-specific and more general limitations of various estimation methods. For example, in retrospective studies, it is difficult to isolate the contribution of seismic monitoring from other factors that influence the reduction in earthquake losses. In theoretical or simulation analyses, it is relatively more difficult to verify the projections of benefits. Although such limitations are not uncommon in many areas of public policy, we can still glean some insights from the few rigorous analyses, from the many studies that yield ballpark estimates, and from theoretical work. Public policy decisions generally have to be made despite such limitations.

Table 3.3 lists methods typically used to estimate the various categories of benefits resulting from reducing earthquake losses. The table summarizes the major benefits from seismic monitoring information and lists the ways in which these methods can be used in the future to more rigorously estimate seismic monitoring benefits.

In applying any of the methods discussed above, it is important to keep in mind several basic principles of benefit-cost analysis. The major ones include:

- Losses must be evaluated in terms of real resource costs and prices that reflect their competitive value. This excludes transfer payments, such as taxes, and may require adjustment in existing prices for various other distortions (e.g., monopoly pricing).
- Benefits are not limited to those activities with markets but should also include nonmarket effects such as externalities (e.g., pollution) or reduction in public goods (e.g., transportation services).
- Future benefits must be discounted to adjust for the "time value of money" (except perhaps in the case of the value of a human life).
- Flow measures of benefits, such as business interruption losses, should be evaluated over the time period during which individual businesses and the economy as a whole have not returned to the projected normal level of economic activity.
- Benefits should not be estimated in a context in which decision-makers are assumed to react passively or in a "business-as-usual" mode to an earthquake; rather, benefits should reflect inherent and adaptive resilience at the individual, market, and community levels (e.g., a signifi-

TABLE 3.3 Possible Methods for Estimating the Benefits of Hazard Loss Reduction Resulting from Improved Seismic Monitoring

Benefit Category	Method
1. Reduced direct property damage	Empirical data Construction cost estimation HAZUS loss estimation Statistical estimation
2. Reduced induced property damage	Empirical data HAZUS loss estimation Statistical estimation
3. Reduced direct business interruption loss	Empirical data HAZUS loss estimation Statistical estimation
4. Reduced indirect business interruption loss	Input-output analysis[a] HAZUS loss estimation Computable general equilibrium[b]
5. Reduced environmental damage	Contingent valuation[c] Hedonic price[d] Benefit transfer[e] Meta-analysis[f]
6. Reduced other nonmarket damage (e.g., historic sites)	Contingent valuation[c] Hedonic price[d] Benefit transfer[e] Meta-analysis[f]
7. Reduced human impacts	Empirical data HAZUS Statistical analysis
8. Reduced government administrative costs	Empirical data
9. Reduced emergency response costs	Empirical data

[a] Linear model of all purchases and sales between sectors of an economy, based on the technological relationships of production (Rose and Miernyk, 1989).

[b] Nonlinear model of the entire economy based on decisions by individual producers and consumers in response to price signals, within limits of available capital, labor, and natural resources (Shoven and Whalley, 1992).

[c] Elicitation of willingness-to-pay statements from survey respondents (Mitchell and Carson, 1989).

[d] Method that bases estimates of benefits on characteristics of an entity itself or opportunity cost of another use, such as the use of housing prices and wage differentials as measures of the implicit price of a wetland or open space (Freeman, 2003).

[e] Method that adapts summary measures of benefits from one study site to another site (Luken et al,, 1992).

[f] Method that uses results of several studies as observations in a synthesis regression analysis (Smith and Pattanayak, 2002).

cant proportion of lost production resulting from electricity outages can be "recaptured" by subsequent overtime work in most sectors).

In the following chapters, the incremental benefits of improved seismic monitoring are discussed in detail. This discussion yields predominantly qualitative benefits, although a lower-bound estimate of the benefits that result from performance-based engineering enhancements is presented in Chapter 6. Why are there not more examples of quantitative benefits, especially given the wide range of potential benefits discussed above? The answer is the lack of certain types of information. Although the committee has argued that a wide array of incremental benefits results from an enhanced seismic network, this argument is forward looking. That is, only over time will these benefits be realized. Just as improved seismic monitoring will provide more critical and basic information to earth scientists and engineers, critical information has to be gathered from future earthquake events and provided to scientists in the behavioral sciences (economics, finance, sociology, political science, psychology, etc.) to fully assess the benefits.

4

Benefits from Improved Earthquake Hazard Assessment and Forecasting

Remarkable advances in our understanding of earthquakes and their effects have occurred during the past century. These advances have been concentrated in the period since 1960, following the installation of a global network of seismic monitoring stations, the World Wide Standardized Seismograph Network (WWSSN). Although the main purpose of the WWSSN was the monitoring of nuclear explosions, the revolution in earth science produced by the acceptance of the theory of plate tectonics in the 1960s could not have occurred without the seismic information provided by this network (a comprehensive summary of these advances was presented in NRC, 2003a). In the context of these advances, this chapter first provides an overview of the role that seismic monitoring plays in hazard assessments in the United States. These hazards include earthquake ground shaking, tsunamis generated by earthquakes, and volcanic eruptions. It then discusses the central role that seismic monitoring plays in the development of ground motion prediction models, which are a vital input into all aspects of earthquake engineering (see Chapter 6). In this context, the committee focuses on a particularly important application of seismic monitoring—the identification of locations within urban regions that are especially vulnerable to damaging earthquake ground motions (seismic zonation). This is followed by a description of the role of seismic monitoring in earthquake forecasting and prediction.

MONITORING FOR HAZARD ASSESSMENT

Each year, tens of thousands of small earthquakes occur throughout the United States, reflecting the brittle deformation of the North American plate along its edges and within its interior. Although not damaging, these smaller earthquakes provide a wealth of information that enables seismologists and engineers to better assess the distribution, frequency, and severity of seismic hazards throughout the country. Seismograph networks supply earthquake parameter and waveform data that are essential for the real-time evaluation of tectonic activity for public safety (e.g., volcanic eruptions, tsunamis, earthquake mainshocks and aftershocks), the development of earthquake hazard maps and seismic design criteria used in building codes and land-use planning decisions (e.g., characterization of seismic sources, ground failure, strong ground motion attenuation), and basic scientific and engineering research.

National and regional earthquake hazard maps published by the U.S. Geological Survey (USGS) and state geological surveys involve the collection and integration of seismograph network data with other geologic and geophysical data, including paleoearthquake chronologies, locations of active faults, determinations of three-dimensional velocity and geologic structure, and wave propagation and attenuation parameters. These earthquake hazard data and maps help define the level of earthquake risk throughout the United States and provide input to risk management decisions at both the national and local levels (see Chapter 2). Efforts to reduce the uncertainties in these data help to clarify the level of seismic hazard and risk and to identify the appropriate mitigation and response strategies for different parts of the country.

Earthquake Monitoring

Seismic monitoring provides a wealth of critical information for earthquake hazard assessment and for improved understanding of the earthquake process. The basic product of earthquake monitoring is the seismicity catalog, a listing of all earthquakes, explosions, and other seismic disturbances (both natural and manmade). Parametric data, such as earthquake origin times, locations, and magnitudes, are used to characterize the frequency and size of earthquakes in a particular region and help identify active faults. Earthquake catalogs play a key role in probabilistic seismic hazard assessment, especially in the eastern and central United States where there is generally insufficient detailed information on active faults and their tectonic causes (NRC, 1996).

The importance of earthquake catalogs for seismic hazard assessment underscores the need for consistent and reliable long-term recording and reporting to reduce uncertainties in magnitude, frequency, and location of earthquakes. The level of earthquake catalog completeness and location accuracy varies as a function of time and location throughout the United States. In northern California, for example, regional catalogs are thought to include every earthquake greater than magnitude 5 since about 1850 and every event of magnitude 4 and greater since about 1880. Location accuracies of pre-instrumental earthquakes are within ±50 km, based on newspaper reports and personal accounts. Instrumentally recorded earthquakes of magnitude 3 and larger began appearing in the catalog in the 1940s. With the addition of ~200 U.S. Geological Survey seismographic stations in the greater San Francisco Bay area—from the late 1960s to the early 1980s (USGS, 2003a)—station coverage and velocity models were sufficient to record earthquakes with magnitudes of 1 to 2 and resulted in location and depth uncertainties being reduced to less than 5 km. Currently, relative epicentral locations are estimated to be better than ±0.5 km (depth uncertainty of ±1 km) for earthquakes within the densest portions of seismic monitoring networks in California (Hill et al., 1990).

Elsewhere in the United States, efforts are under way by the USGS, through the Advanced National Seismic System (ANSS) program, to provide uniform coverage in areas not currently monitored by regional networks. Regional network operators throughout the United States have begun to implement ShakeMap capabilities in cities such as Portland, Oregon; Reno, Nevada; and Salt Lake City, Utah. The benefit of this capability was demonstrated in Seattle, Washington, where the first deployment of ANSS instruments was made a few months prior to the 2001 Nisqually earthquake (see Figure 2.1). In areas where sufficient instrumentation is still lacking, such as the central and northeastern United States, it is only possible to issue model rather than observational or empirical ShakeMaps.

Variations in network configurations with time, due to changes in instrumentation and sensitivities as well as changes in procedures for computing earthquake magnitudes and locations, introduce additional uncertainty. Improving the completeness and accuracy of these catalogs is a major objective of seismic hazard analysis, often depending on the occurrence of small earthquakes to identify the potential for damaging fault ruptures, and of earthquake physics, which relies on catalogs as the basic space-time record of fault system behavior.

Changes in technology during the last two decades have resulted in improvements in the detail, quality, and usefulness of seismic data through the deployment of three-component digital and strong motion sensors capable of reliable on-scale recordings over a range of earthquake sizes. The majority of strong motion networks in the United States were established prior to the ready availability of digital technology and,

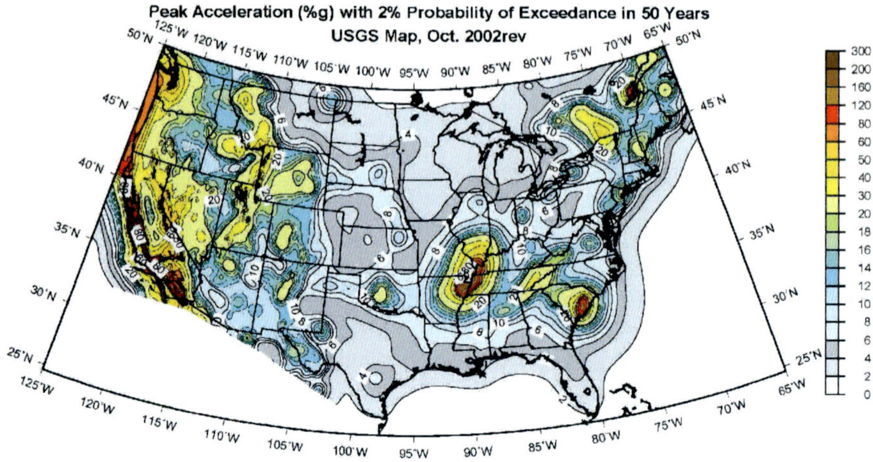

FIGURE 4.1 Seismic hazard map for the United States, showing the distribution of ground motions (in %g) with a 2 percent probability of exceedance in 50 years or a 2,475-year return period.
SOURCE: Frankel et al., 2002.

although still useful, do not meet the current needs of engineers and emergency management officials. As of 1999, only 6 percent of the operating seismographs in the United States could accurately record both very small and fairly large earthquakes on-scale (USGS, 1999).

Digital waveform data, either weak or strong motion, are used to further improve earthquake locations, characterize seismic source and wave propagation effects, measure the state of stress in the brittle crust, and develop ground motion attenuation models. Digital waveform data also have many uses in earthquake engineering, as described in Chapter 6.

Monitoring in the Urban Environment

For close to a century, standard seismological practice has been to site delicate instruments far from urban centers and other sources of noise. Studies of weak ground motions, faint vibrations from earthquakes occurring around the globe, led to important scientific advances during the twentieth century. Recent earthquakes, however, have dramatically demonstrated the vulnerability of the urban environment to earthquake-related damage. Unprecedented growth in urban areas during the last few decades has served to increase the level of earthquake risk faster than our efforts to reduce or mitigate it. Addressing seismic hazard and risk issues in the urban environment has required a change in the standard seismological practice, with the recognition that instruments have to be installed in cities to record ground motions where the earthquake damage is occurring. Recording on-scale ground motions close to active faulting (the near field) and within structures, and obtaining a better understanding of ground response in urban areas, have become critical elements in the national goal of reducing seismic risk. The existing ground motion hazard maps—as illustrated in Figure 4.1—provide information on a national scale, and these will increasingly have to be supplemented by more detailed maps representing seismic zonation for urban areas.

Tsunami Monitoring

Tsunamis are oceanic gravity waves that may be caused by submarine earthquakes or other geologic processes such as volcanic eruptions or landslides. In the United States, tsunamis present a significant (although relatively infrequent) danger to coastal communities in California, Washington, Oregon, Alaska, Hawaii, and Puerto Rico. Seismic monitoring to detect large subduction zone earthquakes around the circum-Pacific and Caribbean regions provides valuable public safety information in advance of tsunami arrivals.

Distant tsunamis and locally generated tsunamis require responses at significantly different time scales. For local tsunamis, the ability to warn coastal communities of a potentially dangerous situation immediately after a large local earthquake is the key to public safety. Locally generated tsunamis can reach the shoreline quickly (within as little as 5 minutes), giving authorities limited time to issue any warnings or evacuations. The 1992 *Mw* 7.1 Cape Mendocino, California, earthquake generated a small 1-foot tsunami that reached Humboldt Bay 20 minutes after the earthquake occurred. Regional groups throughout the Pacific Northwest—such as the University of Washington, the Oregon Department of Geology and Mineral Industries, and the Bonneville Power Authority—recognize the significant local tsunami hazard posed by the Cascadia subduction zone and have begun installing strong motion instruments for real-time monitoring and warning for coastal communities.

Distant or tele-tsunamis generated from other parts of the circum-Pacific are monitored by the Pacific Tsunami Warning System, which was established in 1948 following the 1946 Aleutian (Unimak Island) tsunami. Tsunami waves travel at speeds of 800 km/h (or 0.2 km/s) at a water depth of 5,000 meters, far slower than seismic waves (3-8 km/s). This difference in wave speed makes it possible to issue tsunami warnings throughout the Pacific basin after an earthquake has been detected, but before the arrival of the tsunami. A Tsunami Watch Bulletin is released when an earthquake occurs with a magnitude of 6.75 or greater on the

Richter scale. A Tsunami Warning Bulletin is released when information from tidal stations indicates that a potentially destructive tsunami exists.

Although great strides have been made over the past 50 years in tsunami detection and warning, 75 percent of all tsunami warnings issued since 1948 were false alarms and evacuation was not required. The cost of evacuation to the Hawaii Gross State Product is estimated to be ~$58 million (1996 dollars) per day (Iboshi, 1996). Not only are these false alarm evacuations costly, they also erode the credibility of the tsunami warning system. Furthermore, the fear and disruption of a false alarm can itself put a population at physical risk—fatalities and injuries have occurred during an evacuation due to such things as heart attacks and accidents.

As part of the National Tsunami Hazard Mitigation Program, the National Oceanic and Atmospheric Administration (NOAA), in cooperation with the USGS and the States of Alaska, California, Hawaii, Oregon, and Washington, is expanding, integrating, and upgrading the network of seismic stations to improve tsunami warnings, reduce false alarms, and better track the source and type of earthquakes for NOAA's tsunami warning centers at Ewa Beach, Hawaii, and Palmer, Alaska. Real-time determination of earthquake source parameters by digital seismograph networks enables faster response times for the warning centers and recognition of "tsunami earthquakes"—events that excite tsunamis that are larger than expected for their magnitude (such as the 1946 Aleutian and 1992 Nicaragua tsunamis). Improved seismic data, coupled with information from deep-sea buoys that detect water pressure changes, will enable the accurate determination of tsunami size in real time and eliminate or reduce unnecessary coastal evacuations. The recent provision of funding to enable the USGS to expand seismic instrumentation for tsunami warning and response,[1] following the Indian Ocean tsunami of 2004, represents an explicit recognition by Congress of the value of seismic networks for emergency response.

Volcano Monitoring

Nearly every recorded volcanic eruption has been preceded by an increase in earthquake activity beneath or near the volcano. For this reason, seismic monitoring has become one of the most useful tools for eruption forecasting and monitoring (McNutt, 2002). Systematic volcano monitoring enabled the accurate prediction, from hours to even a few weeks in advance, of nearly all the post-May 18, 1980, dome-building

[1]H.R. 1268: *Emergency Supplemental Appropriations Act for Defense, the Global War on Terror, and Tsunami Relief, 2005.*

eruptions of Mount St. Helens. Real-time and near-real-time seismic monitoring capabilities at numerous volcanoes around the world provide a major advance for identifying and guarding against volcano hazards. In addition to monitoring, the improved ability to locate earthquakes recorded by permanent seismic networks provides three-dimensional images of the magmatic plumbing systems beneath some volcanoes. The increasing use of broadband seismometers has facilitated the complete recording and comprehensive analysis of long-period seismic signals, which have preceded and accompanied a number of eruptions. A more quantitative understanding of long-period seismicity not only refines short-term forecasts of volcano hazards, but also improves our knowledge of magma transport and eruption dynamics.

The economic consequences of volcanism in the United States are wide and varied, ranging from the destruction associated with the May 1980 eruption of Mount St. Helens, Washington (~$1 billion in losses and 57 fatalities), to the impacts on air transportation from high-altitude ash clouds, to fluctuations in real estate values as a societal response to official warnings (e.g., Mammoth Mountain-Long Valley Caldera, California, earthquake swarms in the 1980s).

In the past 30 years, more than 90 jet-powered commercial airplanes worldwide have encountered clouds of volcanic ash and suffered varying amounts of damage as a result (Guffanti and Miller, 2002). The overall economic risk from airborne volcanic ash effects is estimated to be about $70 million per year (Kite-Powell, 2001). More than 10,000 passengers and millions of dollars of cargo fly across the North Pacific region each day, and the area's aviation traffic is increasing at a rate of 10 percent per year (USGS, 2004). Coordinated observations, using both land- and space-based data, are needed to evaluate volcanic threats in real time. Seismic monitoring coupled with satellite observations and ash-cloud transport models enables the air transportation industry to reroute flights and avoid costly ash-cloud encounters. More than 100 potentially dangerous volcanoes lie under air routes in the North Pacific. Along the Alaska Peninsula and the Aleutian Islands there are more than 41 historically active volcanoes. As of July 2002, the Alaska Volcano Observatory operated networks at 23 of the most dangerous volcanoes in Alaska and had plans to instrument additional volcanoes to achieve the ANSS goal of having all potentially active volcanoes in the United States monitored by at least three seismograph stations within 20 km of the volcano.

MONITORING FOR GROUND MOTION PREDICTION MODELS

Earthquake engineering practice uses ground motion prediction models to estimate ground motion levels for the design of structures. These models predict the level of the ground motion of future earthquakes based on the earthquake magnitude, the distance of the site from the earthquake, and the nature of the shallow surface geology (soil or rock) at the site. In the western United States, ground motion models are based mostly on the recorded ground motions of past earthquakes. In the central and eastern United States, they are based mostly on computer simulations of earthquake ground motions derived using seismological theory (e.g., Abrahamson and Shedlock, 1997).

The strength of earthquake ground motion has a large degree of variability from one location to the next, even when these locations are at the same distance from the same earthquake. For this reason, ground motion models specify two measures of the ground motion level—the average value and the variability (standard deviation) about this average value (e.g., Figure 4.2). The standard deviation in ground motion models typically has values that range from a factor of 1.5 to a factor of 2, and the total range of variation can approach a factor of 10. This large degree of variability is reflected in the irregular distribution of both damage and ground shaking intensity patterns observed following earthquakes.

Use of Ground Motion Prediction Models for Building Codes and Seismic Design

Because of the uncertainty in the location and timing of future earthquakes, engineers generally take a probabilistic approach to characterizing the strength of future earthquake ground motions for seismic design at a given site. This probabilistic approach is at the core of current building code and seismic design practice (ICC, 2000; FEMA, 2001b). In this approach, the frequency with which a given ground motion level is expected to occur at a site is calculated based on consideration of the frequency of occurrence of all of the possible earthquakes that could occur on all of the faults that are close enough to affect the site. The probabilistic ground motion calculation also takes into account the variability in the level of the ground motion expected from a given earthquake.

In the FEMA (2001b, 2001c) National Earthquake Hazards Reduction Program (NEHRP) seismic provisions, which form the basis of current building codes, seismic design at most locations is based on the ground motion level that has a 1/2,500 chance per year of being exceeded. In seismically active areas such as the coastal regions of California, earthquakes recur on some faults as frequently as once every few hundred years. Consequently, construction design may have to accommodate the largest ground motion that would be expected from the occurrence of 10 earthquakes on such faults. If there were no variability in the ground motion level caused by a given earthquake, then the largest ground

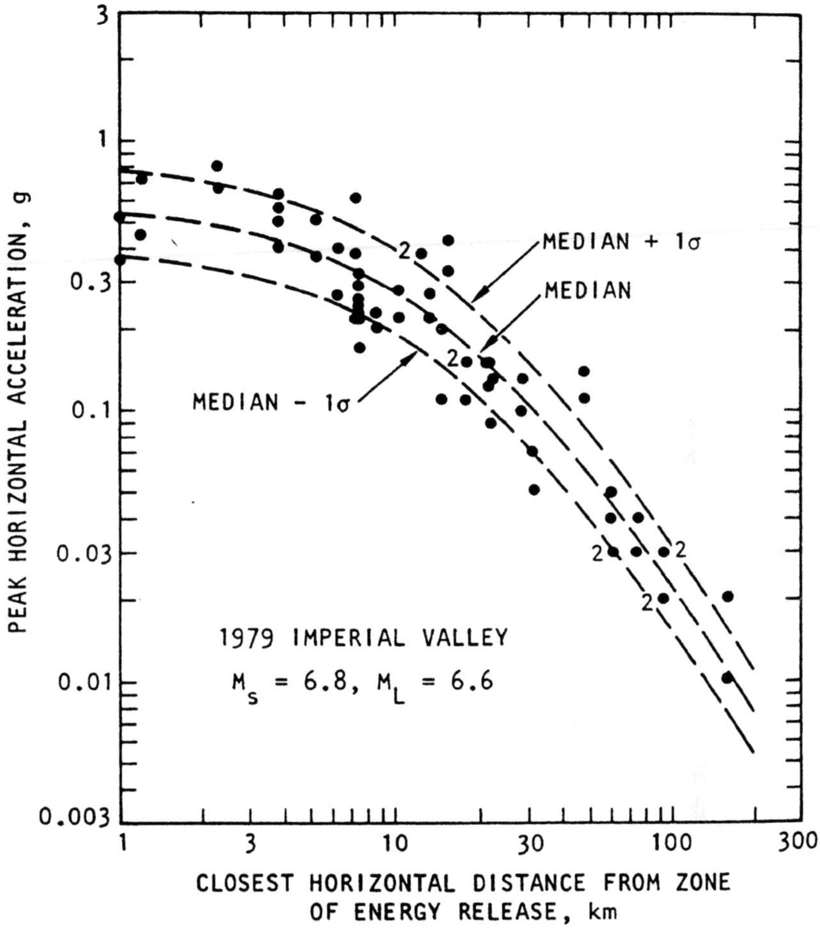

FIGURE 4.2 Recorded peak accelerations of the 1979 Imperial Valley earthquake and an attenuation curve that has been fitted to the data.
SOURCE: Seed and Idriss (1982).

motion level from the occurrence of 10 earthquakes would be the same as that from a single occurrence—the average value. However, given the variability in the ground motion level, we would expect the largest ground motion level from the occurrence of 10 earthquakes to be higher than that from one, because with 10 earthquakes there is a higher probability that one of them would produce a ground motion level that is much higher than the average value for that earthquake.

The effect of the variability in ground motion is to greatly increase the probabilistic estimates of ground motion levels at annual probabilities of 1/2,500. Near large, active faults in the western United States, the probabilistic ground motion levels are so high that engineers have decided not to use them in building codes, and instead they apply a cap to the ground motion levels used in the building code (FEMA, 2001b, 2001c). They have done this because the damage that they have observed in earthquakes does not seem to them to be as great as would be expected if the actual ground motions were as large as the values that are projected to occur based on current ground motion models and structural analysis techniques. This situation results from a fundamental lack of data that can be used to describe the strength of earthquake ground motions. Until the ground motion recordings that can resolve this issue are obtained, the ground motion levels used in building codes will continue to be based more on professional judgment than on real knowledge of ground motion characteristics. This lack of adequate strong motion recordings is hindering the continued development of performance-based design (see Chapter 6), whose goal is to develop reliable procedures for predicting the performance of structures when subjected to earthquake ground motions. This kind of predictive capability is necessary to provide a more rational basis for the cost-effective design of structures.

The Need to Record Damaging Ground Motions

The current seismic monitoring networks in the United States are designed mainly to detect earthquakes and, for that purpose, use very sensitive transducers that can record the smallest possible earthquake magnitudes. When an earthquake occurs that is large enough to be damaging (with accelerations more than about 10 percent of the acceleration of gravity), most of the nearby seismic recording instruments are driven off scale. This results in the loss of information about how strongly the ground was shaken at the sites of damaged structures, making it difficult for engineers to assess building performance relative to actual ground motions. One of the main purposes of the ANSS program is to rectify this situation by installing instruments that will remain on scale during damaging earthquake ground motion. These instruments will be installed both on the ground, to measure the strength of the shaking that enters structures, and within structures, to measure the behavior of the structures in response to ground shaking. The ground instruments will have the capability of recording both strong ground motion (i.e., motions having potentially damaging levels, directly important for earthquake engineering) and weak ground motion (i.e., barely perceptible motions from small or distant earthquakes, directly applicable to earthquake monitoring and earthquake hazard assessment).

Paucity of Design-Level Strong Motion Recordings in the Western United States

There is a large and growing data set of strong motion recordings of earthquakes in the western United States, mostly from California. However, there are still very few nearby recordings of the large-magnitude earthquakes that control the design of structures in the western United States. Consequently, many structures are designed for ground motions that are stronger than any that have been recorded in this country. The development of ground motions for engineering design thus involves extrapolation of ground motion models to larger magnitudes and closer distances than are reliably represented by recorded data. The resulting uncertainty in the levels of ground motions that are suitable for seismic design gives rise to the use of conservative assumptions that result in unnecessary expense.

The strong ground motions recorded during recent large earthquakes in other countries, including Turkey, Taiwan, and Japan—the latter two of which have strong motion recording systems that are vastly superior to those in the United States—all point to the likelihood that our current ground motion models are too conservative (Somerville, 2003). Confirmation of this finding from recordings of earthquakes in the United States is essential. The *Mw* 7.9 earthquake that occurred in Alaska in 2002 presented such an opportunity, occurring on a well-known active fault (the Denali fault) and causing a rupture length of almost 400 km. However, there was only one strong motion recording near the fault, made by the Alyeska Pipeline Company close to the location where the oil pipeline crosses the fault. If the Denali fault had been adequately monitored with recording instruments, a valuable data set would have been recorded for use in seismic design in the western United States.

Paucity of Strong Motion Recordings in the Eastern and Central United States

At present, there are very few strong motion recordings of earthquakes in the eastern and central United States, both because the level of seismic activity is quite low and because very few strong motion recording instruments are located in this region. When an earthquake does occur in this region, most of the closest recording instruments go off scale, and as a result there are insufficient strong motion recordings to develop ground motion models. Until there is adequate monitoring of strong ground motion in the central and eastern United States, earthquake engineering design in this region will continue to be subject to very large uncertainty and the concomitant economic effects of potential "underdesign" (resulting in high damage levels) or "overdesign" (resulting in needless construction costs).

Because of the paucity of strong motion recordings available in the central and eastern United States, ground motion models for this region are based mostly on computer simulations of earthquake ground motions derived using seismological theory. These theory-based models themselves are based on information on the characteristics of the earthquake source and of seismic wave propagation through the earth. As described earlier, the information about these characteristics is derived from the modeling and analysis of seismograms recorded on seismic networks. To a large extent, these analyses can use information from quite small earthquakes, which occur with sufficient frequency to provide useful results. These small earthquakes do not themselves generate strong motions, but the weak motions that they generate can be used to understand earthquake source and wave propagation characteristics, which can then be extrapolated for the prediction of strong ground motions from larger earthquakes. This means that while we are waiting for strong ground motions to be recorded from future large earthquakes in the central and eastern United States, recordings from smaller earthquakes can be used to improve our current theory-based ground motion models.

SEISMIC ZONATION FOR REDUCING UNCERTAINTY

As noted above, the pattern of damage caused by earthquakes often has a highly irregular distribution, with concentrations of damage in some locations and relatively little damage in others. In communities where strong motion recordings are available, these spatial variations in damage characteristics are usually reflected in a general way by the distribution of recorded ground motions of the mainshock (e.g., as depicted by ShakeMap) and by the spatial variations of weak ground motions recorded during aftershocks. The ability to reliably predict the pattern of ground motion amplification in urban areas, and thus identify locations that are especially vulnerable as well as those that are not, has the potential to significantly reduce earthquake losses and guide rational urban development. Box 4.1 demonstrates how improved information about the location and levels of ground shaking impacts bridge design and construction costs in California. However, the development of this capability is contingent upon the deployment of dense arrays of strong motion recording instruments in urban regions, as planned for ANSS.

BOX 4.1
Seismic Zonation and Bridge Construction Costs

Transportation systems are examples of geographically distributed networks. Figure 4.3 shows levels of ground amplification in southern California, where a number of transportation corridors cross areas with the highest levels of ground amplification. Increased specificity (reduced uncertainty) as to both the location and the level of ground shaking enables geographically optimized risk management and mitigation decisions for both the long and short term. Long-term benefits are achieved through improved and more appropriate design and construction. Improved information about ground motion demand impacts $10 million to $60 million of new bridge construction costs each year in California (based on $1 billion spent each year on bridges in California). Figure 4.4 shows that a 10 percent change in ground motion results in a 6 to 15 percent change in construction costs.

Short-term benefits are realized during the earthquake response and recovery phase. The California Transportation Department (CalTrans) and (CSMIP) have invested ~$7million in strong motion monitoring of bridges. Information about the severity of ground shaking at specific bridge locations and the capacity of those bridges allows CalTrans to prioritize post-earthquake inspection and repair activities. Reduced traffic delays through efficient rerouting impact both the immediate emergency response and the

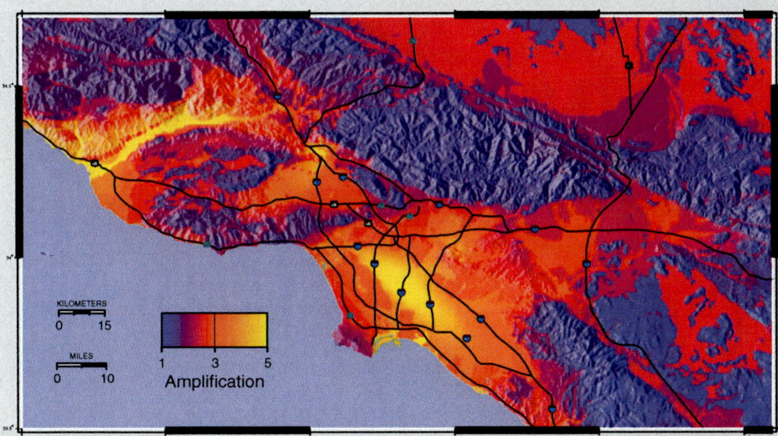

FIGURE 4.3 Map showing earthquake ground motion amplification in Southern California.
SOURCE: Field (2001).

continued

BOX 4.1 Continued

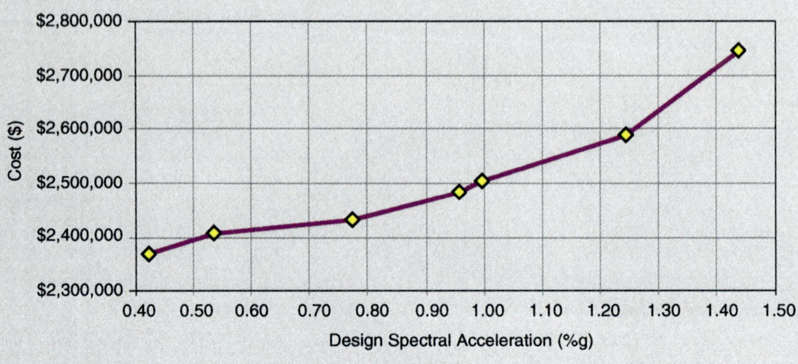

FIGURE 4.4 Cost function graph, showing increased construction costs as a function of design ground motions.
SOURCE: Ketchum et al. (2004).

longer-term recovery phases. Costs associated with loss of access are estimated to be up to $1 million per day per route.[a]

[a]Presentation to committee meeting by C. Roblee, December 16, 2003, San Francisco, California: *Transportation Perspective: Benefits of Improved Seismic Monitoring.*

Causes of Spatial Variations in Ground Motion Levels

In some cases, the spatial variations in ground motions can be attributed to spatial variations in the near-surface geology. Indeed, building code provisions include the effect of the shallow geology (specifically, V_s, the average shear wave velocity in the upper 30 meters) on ground motion amplitudes. Accordingly, mapping of V_s in urban regions can be used as a first-order method of seismic zonation. As with all methods of seismic zonation, the use of V_s measurements to quantify site amplification requires experimental verification. For example, a major field program (ROSRINE)[2] was undertaken following the 1994 Northridge earthquake

[2]See http://geoinfo.usc.edu/rosrine/.

to measure V_s at some of the sites that recorded the earthquake ground motions, for use in testing methods of predicting site response.

However, in many other cases, there remain large spatial variations in ground motion levels that cannot be explained simply by the shallow geology. The large spatial variations in ground motion level in the Seattle area from the 2001 Nisqually earthquake and its aftershocks, and their correlation with damage (Figure 2.1), have already been noted. Another example is provided by the damage pattern caused by the Northridge earthquake, which was characterized by pockets of localized damage that were not clearly correlated with surficial soil conditions (Hartzell et al., 1997). It appears that—in both the Nisqually and Northridge earthquakes—deeper-lying geological structure may have had as much influence on strong motion patterns as the upper 30 meters that are conventionally used to characterize site response.

Basin-Edge Effects. The 1994 Northridge and 1995 Kobe earthquakes showed that large ground response motions may be influenced by the geological structure of fault-controlled basin edges. The largest ground motions in the Los Angeles basin during the Northridge earthquake were recorded just south of the Santa Monica fault. In this region, the basin-edge geology is controlled by the active strand of the Santa Monica fault (Figure 4.5) (Graves et al., 1998). Despite having similar surface geology, sites to the north of the fault (closest to the earthquake source) were subjected to relatively low amplitudes, whereas more distant sites to the south of the fault exhibited significantly larger amplitudes, with an increase in amplification occurring at the fault scarp. This pattern is dramatically reflected in the damage distribution indicated by red-tagged buildings (Figure 4.5), which shows a large concentration of damage immediately south of the fault scarp in Santa Monica. The strong correlation of the ground motion amplification pattern with the fault location indicates that the underlying basin-edge geology controlled the ground motion response, with the large amplification caused by the constructive interference of direct waves with surface waves generated at the basin edge.

The 1995 Kobe earthquake provided further evidence from recorded strong motion data, supported by wave propagation modeling using basin-edge structures, that ground motions may be particularly large at the edges of fault-controlled basins. The Kobe earthquake caused severe damage to buildings in a zone about 30 km long and 1 km wide, and offset about 1 km southeast of the fault on which the earthquake occurred (Pitarka et al., 1998). The basin-edge effect caused a concentration of damage in a narrow zone running parallel to the faults through Kobe and adjacent cities.

Wave Focusing Effects. Although basin-edge effects confirm the role of deep geological structure in causing the local amplification of ground

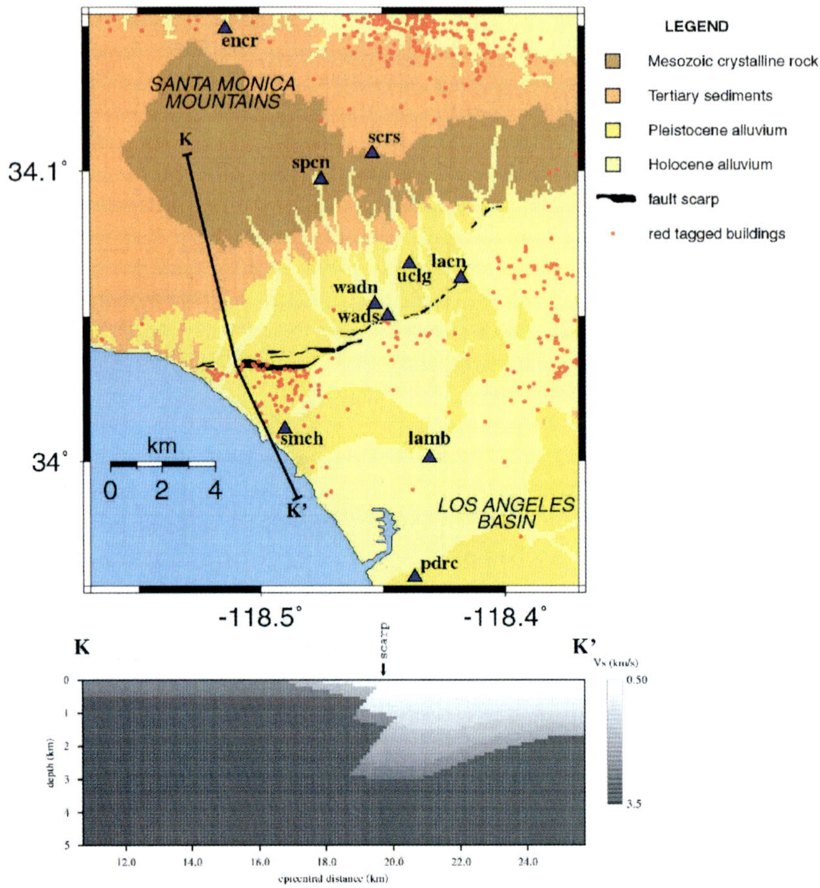

FIGURE 4.5 The surface waves generated in the west Los Angeles basin during the 1994 Northridge earthquake were trapped initially in the shallow sediments north of the ENE-WSW-trending Santa Monica fault. The abrupt deepening of the Los Angeles basin at the Santa Monica fault (shown in light colors on cross section K-K′) caused the basin-edge wave to form a large, long-period pulse of motion that resulted in substantial damage immediately south of the fault, as shown by the distribution of red tagged buildings.
SOURCE: Graves et al. (1998).

motions, in other locations the reason for the localization of damage and ground motion amplification remains obscure. Without detailed knowledge of the deeper structure (provided in the case of Santa Monica by seismic exploration for oil), it is difficult to predict the spatial variation of ground motion levels due to deeper geological structure. Such deeper structure may include structures in the upper few kilometers of sedimentary basins as well as topographic relief of the sediment-basement interface. These structures may focus energy in spatially restricted areas on the surface, in some cases becoming the dominant factor in the modification of local ground motion amplitudes.

Earthquake Source Effects. Recordings of recent large earthquakes have shown that using only the traditional variables of earthquake magnitude, distance from the fault, and site conditions to predict ground motion levels does not adequately describe the observed spatial variations in ground motions. This has motivated the development of ground motion models that use a more complete description of the earthquake source and wave travel path parameters. For example, the long-period pulse of near-fault ground motion caused by forward rupture directivity of the earthquake source (Somerville, 2003), was responsible, together with the basin-edge effect described above, for the intense damage caused by the 1995 Kobe earthquake. Ground motions recorded on the hanging wall above the fault planes that generated the 1994 Northridge and 1999 Chi-Chi, Taiwan, earthquakes were much stronger than the ground motions on the adjacent foot wall, due to the geometrical effects of proximity to the fault. Analyses of strong motion recordings of recent earthquakes such as these are forming a basis for the capability to predict these effects, but many more recordings are needed to develop a comprehensive understanding of these phenomena.

Seismic Zonation of Urban Regions Using ANSS

In most cases, the only way to identify patterns of ground motion amplification in urban regions before the occurrence of a damaging earthquake is to deploy dense urban arrays of strong motion recorders, as planned for the ANSS program (USGS, 1999). These ANSS instruments will contribute to the seismic zonation of urban regions in three ways:

Identifying Zonation by Recording Damaging Motions. The few strong motion instruments located in urban regions in the United States are insufficient to provide an adequate description of the spatial distribution of ground motion. The proposed ANSS instruments will record damaging earthquakes on scale, providing the data that are needed for understanding the distribution of ground shaking level and its relationship to the distribution of damage (e.g., Figure 4.5).

Identifying Zonation by Recording Weak Motions. In most urban regions, damaging earthquakes occur infrequently. However, ANSS monitoring instruments will continuously record weak ground motions from the more frequent, smaller earthquakes. The pattern of ground motion amplitudes from these small earthquakes can potentially provide useful information about the likely distribution of damaging ground motions that would occur in strong earthquakes. The recording of aftershocks following a large earthquake, often using portable deployments of instruments, is used for the same purpose of mapping the spatial variations in ground motion amplification in urban regions (Figure 2.1).

Identifying Zonation by Developing Predictive Capabilities. The ground motions recorded by seismic monitoring instruments from both small and large earthquakes provide data that can be used to identify the deep geological structure beneath urban regions. The adequacy of crustal structure models can then be tested to assess whether seismological ground motion simulation techniques are able to predict the observed patterns of ground motion amplitudes. If they are, then the models can be used to produce ShakeMaps for future scenario earthquakes. The ground motion amplification patterns in these ShakeMaps can then be used to develop a seismic zonation of the urban region, providing a basis for the prioritization of earthquake risk mitigation activities.

MONITORING FOR EARTHQUAKE FORECASTING, ALERTS, AND PREDICTION

Uncertainties about the timing and magnitude of future earthquakes have led seismologists to adopt a probabilistic approach to describing the likelihood of future damaging events. Forecasts, which may involve low probabilities, are distinguished from predictions, which involve probabilities that are high enough to warrant public response. Accordingly, prediction refers to situations in which the probability of occurrence of an earthquake is much higher than normal in a specified region.

Earthquake alerts can take several forms. Short-term (24-hour) forecasts of earthquake hazard, updated every hour, have recently been implemented throughout California.[3] Following a large earthquake, the population may be alerted to the likelihood of aftershocks or the possibility that an even larger earthquake might occur (Jones and Reasenberg, 1989; Reasenberg and Jones, 1994). Alerts may also be issued when increased seismic activity in a seismic "hot spot" is interpreted as a possible precursor to a larger event. In the few seconds to tens of seconds immediately following an earthquake, advanced seismic monitoring systems can provide warning of the imminent occurrence of strong ground shaking—the role of immediate alerts (real-time earthquake warnings) for emergency management is discussed in Chapter 7.

---

[3]See http://pasadena.wr.usgs.gov/step/.

Earthquake prediction is commonly understood to mean specification of the location, time and magnitude of an impending earthquake within specified ranges of uncertainty (Allen, 1976). Earthquake predictions can be divided into short-term predictions (hours to weeks), intermediate-term predictions (1 month to 10 years), long-term predictions or forecasts (10 to 30 years), and long-term potential (>30 years) (Sykes et al., 1999). Since an earthquake prediction might be fulfilled by chance, there is general agreement that probabilistic methods should be used to evaluate the success of any earthquake prediction. The development of a reliable earthquake prediction capability could potentially provide the means to move populations out of harm's way, prioritize the retrofitting of seismically vulnerable infrastructure, and modify urban planning to minimize earthquake risk. The role that seismic monitoring plays in predictions and forecasts is addressed in the following sections.

Long- and Intermediate-Term Forecasts and Predictions

H.F. Reid's (1910) observation that the 1906 San Francisco earthquake was the result of a sudden relaxation of elastic strains through rupture of the San Andreas fault laid the foundation for the development of elastic rebound theory and long-term earthquake forecasting. Large earthquakes are thought to occur more or less regularly in space along major fault systems and, in time, as a result of gradual stress buildup and sudden release by failure. This repetitive cycle of strain accumulation and release, termed the seismic or earthquake cycle (Scholz, 1990), is driven by plate tectonics along the world's major plate boundaries and fault systems.

The seismic gap method is a frequently used application of this concept for long-term earthquake forecasting and the identification of seismic potential. Along many simple plate boundaries, like the San Andreas Fault, most of the long-term plate motion occurs during infrequent large and great earthquakes. As originally proposed by Kelleher et al. (1973), sections of active plate boundaries that have not been the site of large or damaging earthquakes for more than 30 years are considered the likely site for future events. This approach was successfully applied for several large (Mw >7.5) earthquakes along subduction zones and strike-slip plate boundaries during the 1960s and 1970s (Fedotov, 1965; Mogi, 1968; Kelleher et al., 1973). Application of this method for events with Mw <7.5, or in areas with complex tectonic settings, is more controversial (Jackson and Kagan, 1991, 1993; Nishenko and Sykes, 1993). During the 1980s and 1990s, the seismic gap methodology was extended to include the characteristic earthquake model (similar sized events that occur repeatedly along the same section of a plate boundary or fault zone) in a probabilistic framework (Schwartz and

Coppersmith, 1984; Sykes and Nishenko, 1984; WGCEP, 1988, 1990, 1995, 1999; Nishenko, 1991).

The repetitive pattern of strain accumulation and release is also expressed in seismicity patterns, where low levels of seismicity in the first part of the cycle (once aftershocks from the latest event subside) are followed by an increase in regional activity as strain reaccumulates and, ultimately, by the occurrence of another earthquake with its attendant foreshocks and aftershocks. Variations in the historic rate of moderate to large earthquakes in the San Francisco Bay area in the decades before and after the 1906 earthquake (Ellsworth et al., 1981) are similar to those described by Fedotov (1965) and Mogi (1968) for the earthquake cycle associated with great subduction zone earthquakes in Japan, the Kuriles, and Kamchatka.

Three examples of long- and intermediate-term earthquake forecasts in the United States—based on applications of elastic rebound theory and the seismic gap method—include the Parkfield, California, earthquake prediction experiment; the 1989 Loma Prieta, California, earthquake; and the 2002 San Francisco Bay area earthquake forecast. Experience with long- and intermediate-term predictions based on simplified or basic models of earthquake occurrence, however, has shown that accurate forecasts can be difficult even for plate boundaries that have seemingly regular historical sequences of earthquakes, as the following example demonstrates.

Parkfield, California. Moderate-sized (Mw ~6) earthquakes have occurred on the San Andreas Fault near Parkfield, California, in 1922, 1934, and 1966, and earlier events in 1857, 1881, and 1901 were also believed to have occurred in approximately the same area. Similarities in size, location, and even waveforms (see Figure 4.6) led seismologists to believe that these were examples of "characteristic" events that have occurred repeatedly along the same section of the San Andreas Fault. Even though there was significant variation in the times between these events, the relatively short average recurrence time of 22 years suggested that the next characteristic Parkfield earthquake would occur in 1988 (1966 + 22 years) and that the probability of that event occurring before 1993 was 0.95. A focused earthquake prediction experiment (the Parkfield Earthquake Prediction Experiment) began in 1985 (Bakun and McEvilly, 1984; Bakun and Lindh, 1985). The earthquake eventually occurred on September 28, 2004, 38 years after the previous event, without any foreshock or other evident precursory phenomena. Unlike the two previous events—which ruptured from the northwest to the southeast—this event ruptured from the southeast to the northwest, further demonstrating the epistemic and aleatory uncertainties that exist in our understanding of earthquake processes. Nevertheless, valuable strong motion data were recorded by

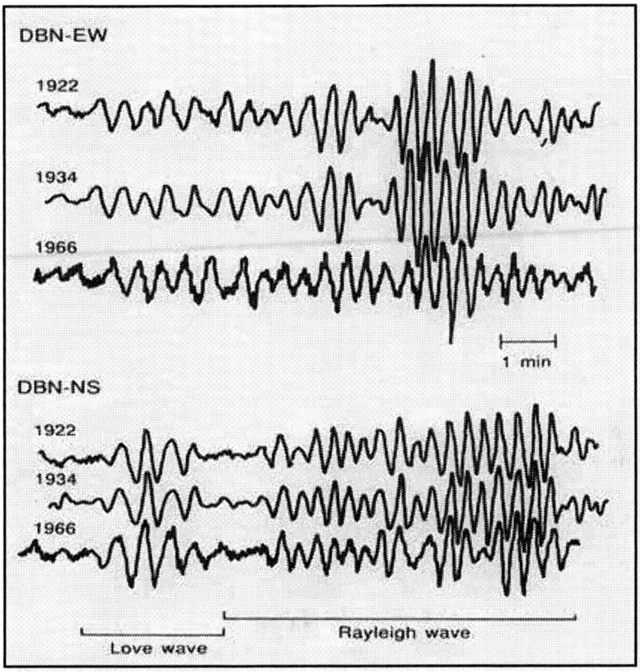

FIGURE 4.6 Comparison of seismograms recorded in DeBilt, Netherlands (DBN), for the 1922, 1933, and 1966 Parkfield, California, earthquakes. SOURCE: Bakun and McEvilly (1984).

arrays of instruments that had been deployed as part of the prediction experiment in the 1980s.

Loma Prieta, California. The 1989 Loma Prieta earthquake (*Mw* 6.9) occurred along an area of the San Andreas Fault where long- or intermediate-term forecasts had been made by a number of seismologists (Lindh, 1983; Sykes and Nishenko, 1984; Scholz, 1985). The Loma Prieta earthquake occurred near the southeastern end of the fault rupture of the 1906 San Francisco earthquake, where the relatively small amount of surface slip in 1906 was thought to have been recovered by elastic strain accumulation. This fault segment also exhibited a distinct absence of small earthquakes over a distance of some 40 km. The general region around Loma Prieta was also identified as having increased likelihood of an earthquake based on pattern recognition studies (see next section). While the 1989 earthquake fulfilled these forecasts in a general sense, details about the earthquake suggest that it may not have occurred on the San Andreas Fault, indicating a greater degree of complexity than was previously recognized for this section of the San Andreas (Harris, 1998).

2002 San Francisco Bay Area Forecast. Building on the lessons learned from earlier California Earthquake Probability Working Groups (WGCEP, 1988, 1990), the USGS revised the 1990 forecast for the San Francisco Bay area and issued a new forecast in 2002. This revised forecast indicates a 62 percent chance of a *Mw* 6.7 or larger earthquake between 2003 and 2032 (USGS, 2003a; see Figure 4.7). These forecasts, like those used in the national seismic hazard maps, are based on the integration of seismograph network data with other geologic and geophysical data, including paleoearthquake chronologies as well as the locations of active faults and their rates of slip. They go beyond the current national seismic hazard maps in also addressing the time-dependent hazard of the region, using information about the time elapsed since the last large earthquake in the region. In contrast, the national seismic hazard maps are time independent, in that they do not change depending on the time from some specific earthquake in the region. These forecasts, coupled with the "wake-up call" from the 1989 Loma Prieta earthquake, have helped to focus earthquake preparedness and mitigation activities throughout the San Francisco Bay area.

Monitoring for Changes in Seismicity. Many earthquake predictions and forecasts have been based on observations of statistically significant changes in the rates and types of seismic activity over long, intermediate, or short time scales. Variations in the state of stress or strength of the crust may be manifest as spatial and temporal changes in seismicity patterns (e.g., "doughnut" patterns, increases or decreases in seismic activity), changes in earthquake focal mechanisms, and increases in the rate of seismic moment release (Mogi, 1969; Wyss and Habermann, 1979; Jaume and Sykes, 1999). The ability to rigorously evaluate the significance of these changes in activity requires a long-term commitment to the systematic collection and evaluation of seismic monitoring data.

Stress Interactions. In the past two decades, it has been recognized that earthquakes frequently occur in locations where the stress has been increased by the recent occurrence of a neighboring earthquake. An earthquake reduces the average value of shear stress on the fault that slipped and redistributes stresses to the fault tips and surrounding regions. Stress trigger zones and stress shadows are regions where previous earthquakes have increased the stress by loading, or decreased it by unloading. The seismicity rate before and after large earthquakes is not constant, and the changes in rate appear to be correlated with these stress trigger zones and shadows. A dramatic example of stress loading is the progressive triggering of 11 large earthquakes along the Anatolian fault system in Turkey between 1939 and 1999 (Stein et al., 1997).

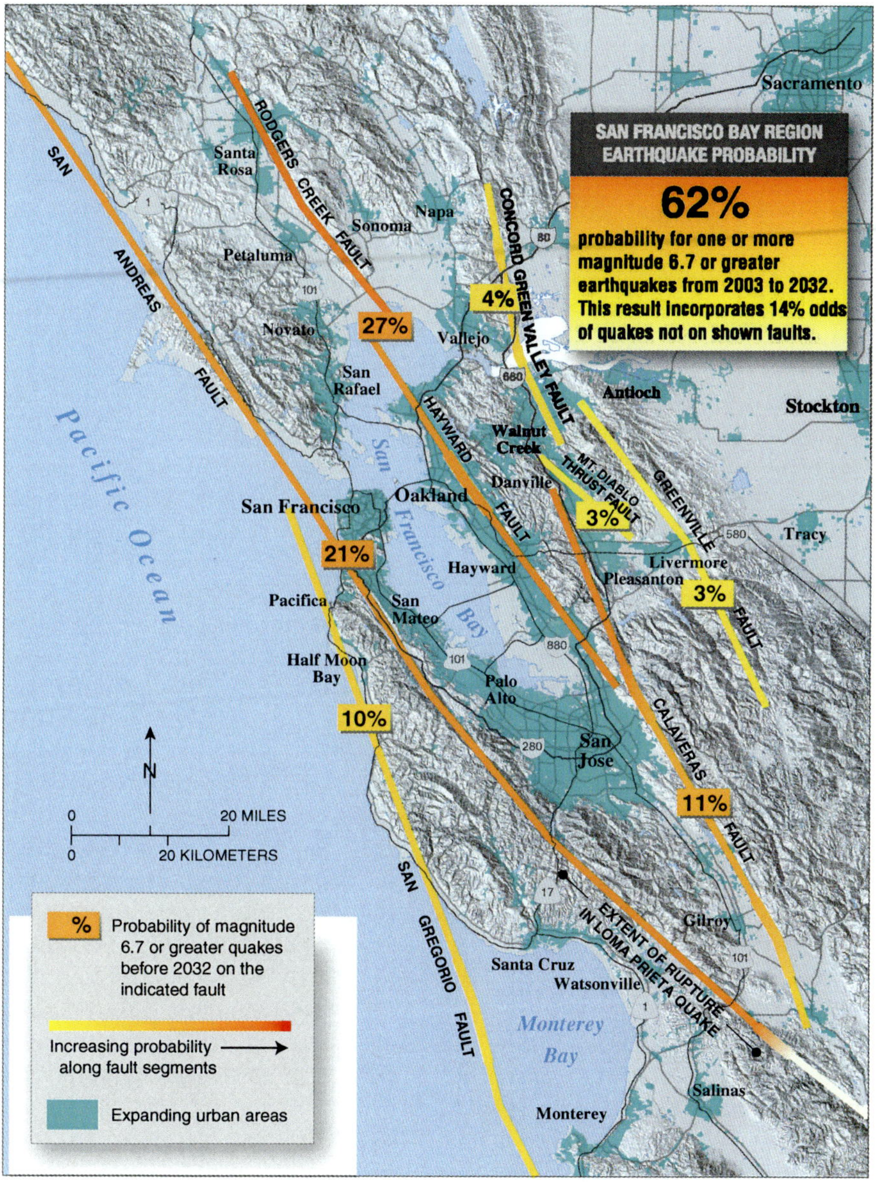

FIGURE 4.7 Long-term earthquake forecast for the San Francisco Bay area. SOURCE: USGS (2003a).

A dramatic example of stress unloading is the stress shadow generated by the 1906 San Francisco earthquake, which decreased the regional seismicity rates in the San Francisco Bay region for the next 75 years (Ellsworth et al., 1981). Earthquake activity was relatively high during the latter half of the nineteenth century leading up to the occurrence of the 1906 event. After 1906, the level of seismicity decreased and moderate-sized events were absent through the first half of the twentieth century. Since the 1950s, the activity level has again increased in the San Francisco Bay area, with the most recent event being the 1989 *Mw* 6.9 Loma Prieta earthquake. While the long-term significance of the increase in regional activity is still being evaluated, the change in seismic activity—punctuated by the 1989 Loma Prieta earthquake—has prompted long-term mitigation and preparedness activities in the Bay area.

There is comparable evidence of seismicity changes in southern California, although the historic record is less reliable until about 1890. Along the rupture zone of the great 1857 Fort Tejon earthquake on the San Andreas Fault, available data show a similar period of low activity for several decades following the event. Farther south, along a section of the San Andreas that has not ruptured since 1630, the activity level since the 1880s is reminiscent of the activity in the San Francisco Bay area in the decades prior to the 1906 earthquake. As a possible long-term indicator of seismic potential, the seismicity surrounding the dormant southern section of the San Andreas agrees with independent assessments of the long-term estimates derived from paleoseismology (Ellsworth, 1990; see also Southern California Earthquake Center [SCEC] Southern California earthquake forecast model in WGCEP, 1995).

Pattern Recognition. Pattern recognition methods are based on statistical changes such as the rate of earthquake occurrence and the proportion of large to small earthquakes within a region. Specifically, the phenomena that are monitored and analyzed include small earthquakes becoming more frequent in an area that is not necessarily where the impending earthquake will occur; earthquakes becoming more clustered in time and space; earthquakes occurring almost simultaneously over large distances within the seismic region; and an increasing ratio of medium-magnitude to small-magnitude earthquakes (Keilis-Borok, 1996). These methods are used to predict the occurrence of a large earthquake within a specified period of time of months to years, in a specified large circular region having a radius of several hundred kilometers (see Box 4.2). The time period of successive predictions has been decreasing as the development of the method has progressed. The developers claim to have predicted a number of earthquakes that occurred within large circular regions: the 1989 *Mw* 6.9 Loma Prieta earthquake within a 5-year window, the 1994 *Mw* 6.7 Northridge earthquake within an 18-month window (although it

BOX 4.2
An Ongoing Earthquake Prediction Experiment

Recent research indicates that, by understanding how stress accumulates in the Earth's crust and how faults interact through the stress changes caused by earthquakes, it may be possible to reduce forecasting times from current intervals of decades to a few years and perhaps even less.

Recent scientific developments have stimulated lots of discussions about the feasibility of what we term "intermediate-term" prediction—forecasting earthquakes on time scales of months to years. Several well-respected groups of geophysicists are running intermediate-term prediction algorithms for research purposes.

For example, last week [January 6, 2004], UCLA issued a press release that a group led by Dr Keilis-Borok had successfully predicted the 25 Sept 03 Hokkaido earthquake (*M* 8.1) and the 22 Dec 03 San Simeon earthquake (*M* 6.5). This release also pointed out that "Keilis-Borok's team now predicts an earthquake of at least magnitude 6.4 by Sept 5, 2004, in a region that includes the southeastern portion of the Mojave Desert, and in an area south of it."

The zone where this event is predicted . . . includes the southern section of the San Andreas, most of the Eastern California Shear Zone south of Interstate 15, the San Jacinto, and several other significant faults.

Dr Keilis-Borok has emphasized the hypothetical nature of this prediction and stressed that his new methodology has not been fully tested. Moreover, the prediction is incomplete, because the authors have not stated the gain in probability over random chance that such an event will occur.

However, other researchers have recognized that the southern San Andreas fault, which lies in the Keilis-Borok prediction zone, is probably late in its seismic cycle and that the seismicity of the region is accelerating.

Despite some discomfort within the scientific community about making predictions that might be misinterpreted by the general public, this type of research must be done in a transparent way and remain open to public scrutiny. Our responsibilities to the public dictate that we maintain up-to-date scientific assessments of earthquake predictability and prediction methodologies, and carefully explore their implications for seismic risk.

At the state level, evaluating earthquake predictions from a public policy perspective is the responsibility of the California Earthquake Prediction Evaluation Council, and it is my understanding that the Keilis-Borok et al. prediction will be put in front of CEPEC for its assessment.[a] There is also the

[a]CEPEC issued a statement on March 2, 2004, that includes the following excerpt: "The Keilis-Borok methodology appears to be a legitimate approach in earthquake prediction research. However, the physical basis for the prediction put forward by the authors has not been substantiated, and they have not yet issued enough predictions to allow a statistical validation of their forecasting methodology."

continued

BOX 4.2 Continued

need for the USGS to reinstate the National Earthquake Prediction Evaluation Council to deal with those issues at the national level.

In conclusion, we can firmly state that the era of time-dependent seismic hazard analysis is at hand. While reliable short-term prediction remains an unreached (and perhaps unattainable) goal, we know earthquake occurrence is inadequately described by assuming earthquakes occur at random times. Many scientists now believe earthquake probabilities have a rich space-time structure that can be quantified if adequate knowledge of the fault system can be leveraged from observations.

Excerpt from address by Tom Jordan, director of the Southern California Earthquake Center: "Progress in Earthquake Science Since Northridge," from *Ten Years Since Northridge: A Special Event for "Movers and Shakers,"* Jan. 16, 2004.

actually occurred 21 days beyond the end of the window), and the 2004 *Mw* 6.5 San Simeon earthquake within a 9-month window. The developers predicted that an earthquake of magnitude 6.4 or larger would occur in southeastern California before September 5, 2004, but no such earthquake occurred.

Short-Term Prediction

An ultimate goal of earthquake science is the short-term prediction of the time, location, and size of an earthquake in a time window that is narrow and reliable enough to enable preparation for its effects. The prediction of the *Mw* 7.3 Haicheng, China, earthquake in 1975, less than 24 hours before its occurrence but in time to allow for evacuation of the population, is credited with saving many lives. However, the prominent foreshocks and hydrological precursors that formed the basis for the prediction were very unusual and were not observed before other earthquakes. Also, many false alarms had been issued, so the possibility of success by chance cannot be ruled out. The limitations of this apparently successful prediction soon became apparent with the occurrence the following year of the *Mw* 7.8 Tangshan earthquake in a neighboring region of China, where the death toll of at least 240,000 is one of the highest in recorded history. This earthquake was not predicted, despite extensive monitoring.

An Alternative View of Earthquake Prediction

Given the definition of earthquake prediction as the specification of the location, time, and magnitude of an impending earthquake within specified ranges of uncertainty, there does not exist anywhere in the world a tested and operational capability to predict earthquakes. However, research that is aimed at developing such a capability is currently under way in several countries. Moreover, operational programs to predict earthquakes, although untested, have been implemented in several locations (including Parkfield, California, and Tokai, Japan, with the latter still in progress), and the pattern recognition method described above is now in a stage of open operation and evaluation. The former two programs are based on the concept of progressing from a long-term forecast of location and magnitude, having very approximate time constraints, to a precise forecast of the time window as data and understanding permit. Lindh (2003) argues that a justification for pursuing earthquake prediction without a tested scientific basis is that society is willing to accept the uncertainties that come with the learning process in which seismologists are engaged because the potential benefits of predicting an earthquake might greatly outweigh the costs of that uncertainty. According to this alternative view of earthquake prediction, earthquakes present such a dire threat that society expects seismologists to attempt earthquake prediction even in the face of great uncertainties, in much the same way as it expects medical doctors to attempt to treat an illness even if its cause is difficult to diagnose.

Prospects for Earthquake Prediction

The following paragraphs describe three promising approaches to earthquake prediction.

1. *Stress interactions.* The stress transfer model for intermediate-term prediction provides a good retrospective explanation for many earthquakes sequences, but it has not yet been implemented as a testable hypothesis because the stress pattern depends on details of the previous earthquake rupture, fault geometry, crustal rheology, fluid flow, and other properties that are difficult to measure in sufficient detail. Monitoring earthquake activity can enhance our knowledge of previous earthquake ruptures, fault geometry, and crustal rheology and thereby enhance the possibility of implementing the stress interaction method as a testable hypothesis for earthquake forecasting.

2. *Pattern recognition.* Pattern recognition methods, which are based on statistical changes in seismic activity within a region, are currently

being tested in a real-time mode (see Box 4.2). Seismic monitoring of earthquake activity can enhance the accuracy of the seismicity on which this method is based and provide physical insight into the tectonic processes that contribute to the phenomena. These include changes in earthquake occurrence rates, earthquake clustering, near-simultaneous occurrence of earthquakes over a wide region, and changes in the proportion of moderate- and small-magnitude earthquakes.

3. *Silent earthquakes.* During the past decade, newly installed dense Global Positioning System (GPS) monitoring systems in the Pacific Northwest and Japan have revealed the occurrence of "silent earthquakes" on subduction plate interfaces. These silent earthquakes are similar to ordinary earthquakes in that they involve the relative movement of one side of a fault past the other, but they differ from ordinary earthquakes in that this slip occurs over a time interval of days to weeks, not the seconds to minutes of fault movement that occurs in ordinary earthquakes. These silent earthquakes are not completely silent—they cause muffled rumblings that are detected by seismic monitoring instruments. These new observations may constitute the means to observe the evolution of deformation on the plate interface that precede the occurrence of an earthquake. For example, a region of the plate interface that has not slipped recently—but lies adjacent to a region that has—may be a candidate for an impending earthquake. Such observations could be the basis for more focused monitoring of particular regions, providing some prospect for the development of an earthquake prediction capability. Regional networks of seismic recording instruments, such as those planned for the ANSS, provide an important component of the earthquake monitoring system needed for the potential development of such a prediction capability.

5

Benefits from Improved Loss Estimation Models

Loss estimation models combine seismic hazard and vulnerability models with inventories of the built environment to estimate the extent of likely damage and the socioeconomic consequences from a range of seismic events. Most of those models are contained in commercial software packages that have been developed by firms specializing in the development and marketing of proprietary models to end users (e.g., the insurance industry). Models in this category include those developed by AIR Worldwide, EQECAT, Risk Management Solutions, and URS. In addition, there are publicly available models, the most widely known and used being the HAZUS model (developed by the Federal Emergency Management Agency [FEMA] and the National Institute of Building Sciences [NIBS]). HAZUS is a standardized, nationally applicable earthquake loss estimation methodology—implemented using PC-based geographic information system (GIS) software—that is intended to be used as a tool for estimating future earthquake losses for the purposes of risk mitigation, emergency preparedness, and disaster recovery.

All the loss estimation models share a common structure. They are based on an estimate of the frequency and severity of the earthquake hazard, coupled with engineering estimates of the damage and loss that would result from events of varying magnitude, applied to the built inventory in a particular region. The inventory typically includes buildings and their contents as well as infrastructure (e.g., roads, bridges, utilities). The output from the models typically includes the amount of expected damage to the built environment, economic costs of that damage (including business interruption costs), and estimates of injuries and deaths. Example outputs are summarized in Table 5.1 for the two most common loss estimation model applications—for insurance (based on proprietary commercial models), response planning, and mitigation (based on publicly available models).

USES OF LOSS ESTIMATION MODELS

Loss estimation models are used by insurers and reinsurers, government agencies, private businesses, the engineering community, and others. Different groups often use the same models and input data but run their analyses for different purposes. Government agencies use loss estimation models during the period immediately after a disaster to help prioritize the allocation of limited resources. Immediately after an earthquake occurs, emergency managers often run a loss model to gauge the scope of the disaster; identify potentially hard-hit areas and localities that may require specialized response (e.g., search and rescue); select locations for staging of emergency resources, shelters, and aid centers (e.g., undamaged areas in close proximity to damaged areas); and accelerate mutual aid requests (see Chapter 7). In non-emergency circumstances, these same loss estimation models are used by emergency managers for exercises to enhance their response plans, by urban and regional planners to identify high-risk areas and to design land-use policies to help mitigate potential losses, and by utilities and public works departments to assess potential infrastructure damage for consideration in their capital improvement plans.

TABLE 5.1 Example Outputs from Loss Estimation Models

Loss Estimation Models for Insurance	Loss Estimation Models for Response Planning and Mitigation (e.g., HAZUS)
• Maximum expected claim cost for a portfolio of insured risks • Expected annual losses (average annual loss) of a specific insured portfolio • Impact of mitigation factors (insurance policy deductibles, improved building features, and other loss avoidance and loss reduction efforts) on expected annual losses and maximum expected claims • Probable maximum loss of a specific insured property (for mortgage securitization)	• Dollar losses associated with damage to buildings (structural, nonstructural, contents, and inventory damage) • Distribution of building damage (damage state) by occupancy and building type • Losses to and post-earthquake functionality of transportation and utility lifelines and essential facilities (e.g., hospitals, schools, police and fire stations) • Regional economic impacts (e.g., direct and indirect business interruption) • Injuries, deaths, and shelter requirements

Similarly, private businesses use loss estimation models in the development of emergency as well as business continuity plans, to facilitate site selection for hazard mitigation and for risk management and insurance decisions. For example, Charles Schwab & Co. in San Francisco has used HAZUS output to develop earthquake planning scenarios—for use in exercises and training—to validate its business continuity plans and to develop products to improve employee awareness.

Insurance and reinsurance companies use loss estimation models to manage and price catastrophe risk. Setting premiums for non-catastrophe coverage is normally based on the loss patterns experienced in the recent past either by the individual insurer or by a group of insurers. The insurance company adjusts the historical information for anticipated changes and uses the results to predict claim costs for the period in which the premiums will be used. Because of the infrequency of catastrophic events, the use of historical information alone provides an inaccurate measure of catastrophic risk. This was never more evident than when both Hurricane Andrew and the Northridge earthquake greatly exceeded insurers and reinsurers estimates of potential losses. Loss estimation models allow the use of all potential events over a long period of time to better estimate the average and long-term impact and to develop worst-case scenarios. This information then allows insurers and reinsurers to better manage the amount of risk they assume—and to better price that risk—to keep the likelihood of a catastrophic loss at an acceptable level.

Engineering professionals use loss estimation models for a variety of purposes, ranging from assessing the effects of proposed mitigation measures on expected building damage during future earthquakes, to helping manage portfolio risk for corporate clients. Standardized loss estimation models can be used for pre- and post-mitigation assessment of potential earthquake damage and loss to individual facilities, portfolios of properties, and geographic regions. The results of such assessments provide critical data for decisions related to individual upgrade design options, proposed mitigation legislation, and future changes to design and construction codes. Many engineering professionals routinely use loss estimation models to help corporate and institutional building owners assess and manage the seismic risk associated with portfolios and individual properties. Decisions concerning how to most cost-effectively reduce (through mitigation), transfer (through insurance), or eliminate (through disposition) seismic risk to a given facility require an accurate assessment of potential earthquake-related losses.

UNCERTAINTY IN LOSS ESTIMATION MODELS

There is inherent uncertainty resulting from the representation of natural processes by computer algorithms. There is additional uncertainty associated specifically with loss estimation models in three primary areas: (1) uncertainty in estimating the likelihood and distribution of ground-motion intensity and ground failure caused by potential earthquakes; (2) uncertainty concerning damage to the built environment caused by the predicted ground motion intensity and ground failure; and (3) uncertainty in the social and economic losses associated with the predicted damage. In the earthquake loss estimation process, these uncertainties tend to be cumulative, often resulting in mean loss estimates with an uncertainty range of at least 2 to 3 times the mean. The problem is exacerbated by the fact that different results are produced by the different private and public loss estimation models—the range of possible results from using multiple models can typically increase the uncertainty of the mean value from a factor of 2 to 3 to more than 4 or 5.

The uncertainty in loss estimation models, including the lack of consistency among the various private and public models, has several impacts on the utility and credibility of the results they produce:

- Uncertainty reduces the credibility of the models with interested parties, particularly when the output contradicts the parties' subjective views of reality.
- Uncertainty decreases the utility of the models for planning purposes, especially when predicted losses from a credible scenario earthquake range from moderate to catastrophic.
- Uncertainty impacts land-use requirements, building code requirements, and building design standards. In some instances, it may lead to unnecessary costs if buildings are designed and built to an unnecessarily strict standard (see Chapter 6). In other instances, buildings are built to inadequate standards due to a lack of credibility associated with the estimates.
- Uncertainty increases the amount of risk associated with the modeled events. For example, insurers, reinsurers, and others assuming financial risk associated with modeled events may demand a higher risk premium as a result of the increased likelihood of unexpected outcomes arising out of uncertainty.
- The high cost of earthquake insurance, resulting in part from the uncertainty associated with estimates of seismic risk, limits the amount of earthquake insurance purchased today. As a result, most homes and businesses in the United States are not insured for earthquake loss, which negatively impacts post-event recovery and increases the demand for government disaster relief.

The primary source of uncertainty in loss estimation models is the lack of accurate input data. This includes not only the data used by the models—such as information about seismic sources, strong ground motion characteristics, local soil conditions, and inventories of the built environment—but also the data used to develop the models driving the loss estimation, such as the relationship between building damage and strong ground motion, and the effects of local soil conditions on the ground shaking intensity. Several studies focusing on sensitivity analysis have illustrated the relative influence on the final loss estimates of the uncertainty associated with each input variable (e.g., Box 5.1). These studies have helped to pinpoint where the additional investment in gathering more accurate data—for inputs to the loss estimation models and for the development of the models themselves—will prove most cost-effective for reducing the uncertainty in the final results and thus improving their credibility and utility.

MONITORING FOR IMPROVED LOSS ESTIMATION MODELS

Improved seismic monitoring is a key element in efforts to reduce uncertainty in loss estimation models resulting from the lack of accurate input data—it will improve loss estimation models in a number of ways:

- Improved seismic monitoring will provide a more complete description of seismic events. This will lead to a better understanding of how different types of faults behave, as well as how seismic energy is transmitted from the source and distributed throughout the impacted region (see Box 5.2).
- Improved seismic monitoring will provide data to improve the models used for estimating how local site conditions contribute to damage, in terms of shaking-induced ground failure and changes in severity of ground motion caused by particular soil types.
- Improved seismic monitoring will increase our understanding of how the built environment is impacted by different levels of seismic activity, primarily by providing the data necessary to develop models that more accurately predict how structures perform during earthquake shaking.

Although some of this information will be captured only in the event of a moderate to large seismic event, there is value to be derived from the improved monitoring of frequent small events, especially in regions of low to moderate seismicity. This "routine" information is valuable for providing a better understanding of how, where, and how frequently seismic events occur, and how the ground motion level attenuates away from the

BOX 5.1
Effect of Uncertainty on Damage Factors

A sensitivity study conducted by Porter et al. (2002) used a tornado diagram (Figure 5.1) to illustrate the impact of the uncertainty in each input variable on the final damage factor for the analyzed building. All input parameters are set to their best-estimate value except for one, which is set to its low (10th percentile) and high (90th percentile) values. The resulting damage factors (ratio of loss to replacement value) are represented by the ends of the horizontal bars. It is important to note that the three most sensitive parameters are those related to earthquake ground motion. *Assembly capacity* refers to the fragility curves (damage models) of the various components of the building that provide estimates of damage as a function of input ground motion; *Sa* is the spectral acceleration of the input ground motion used in the analysis; and *ground motion record* is the specific earthquake record and scaling factor used in the analysis. The remaining parameters—*unit cost, damping, f-d multiplier, mass,* and *O&P*—are related to the repair cost and structural characteristics of the building and are shown to have a smaller influence on the damage factor estimate.

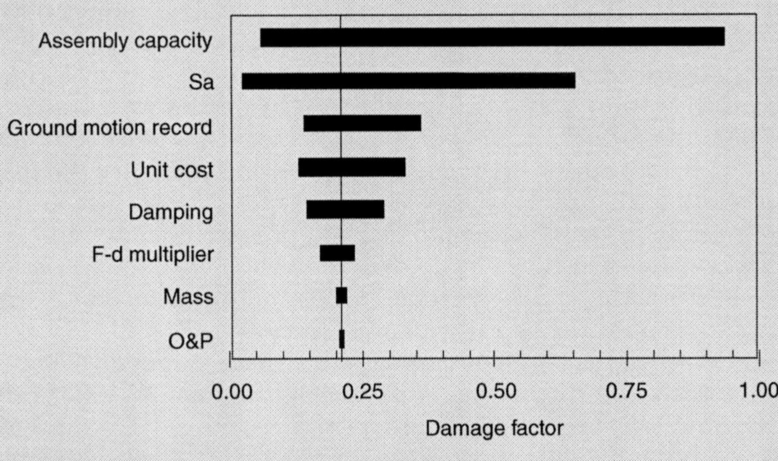

FIGURE 5.1 Results of a sensitivity study describing the relationship between building loss estimates and a number of variables.
SOURCE: Porter et al. (2002).

BOX 5.2
Comparison of Losses, Santa Cruz Mountains

The results from loss estimation models are highly sensitive to the input distribution of ground shaking. Figure 5.2 is a scenario ShakeMap showing the distribution of shaking intensity in the northern California region for a hypothetical *Mw* 7.0 event in the Santa Cruz Mountains. The shaking intensity is symmetrical about the scenario fault rupture, and attenuates evenly with increasing distance from the fault. Loss estimates produced using this scenario shaking distribution are used for earthquake response planning, and—in the absence of seismic monitoring instruments to capture shaking data during an actual event in this region—these estimates would be used for real-time post-earthquake response.

Figure 5.2 can be compared to Figure 5.3, showing the ShakeMap for the shaking intensity in northern California recorded in the 1989 *Mw* 6.9 Loma Prieta earthquake, with an epicenter in the same vicinity as that shown in Figure 5.2. Figure 5.3 shows that the recorded motion is not distributed symmetrically about the epicenter and the motion is significantly higher throughout the region. Loss estimates (and the post-earthquake response decisions based on those estimates) produced from the recorded shaking distribution shown in Figure 5.3 would be drastically different from those produced from the hypothetical shaking distribution shown in Figure 5.2. In fact, using HAZUS-99 (Service Release 2.0), the losses for Santa Cruz County (located in the epicentral region of the earthquakes shown in Figures 5.2 and 5.3) are significantly lower for the scenario (Figure 5.2) than the actual event (Figure 5.3). Tables 5.2 and 5.3 summarize the differences in direct building economic losses and casualties, respectively. Improved seismic monitoring will help improve loss estimation models by providing a more realistic distribution of ground shaking in actual events, and by providing data to develop models that more accurately predict regional distributions of ground shaking for scenario events.

continued

BOX 5.2 Continued

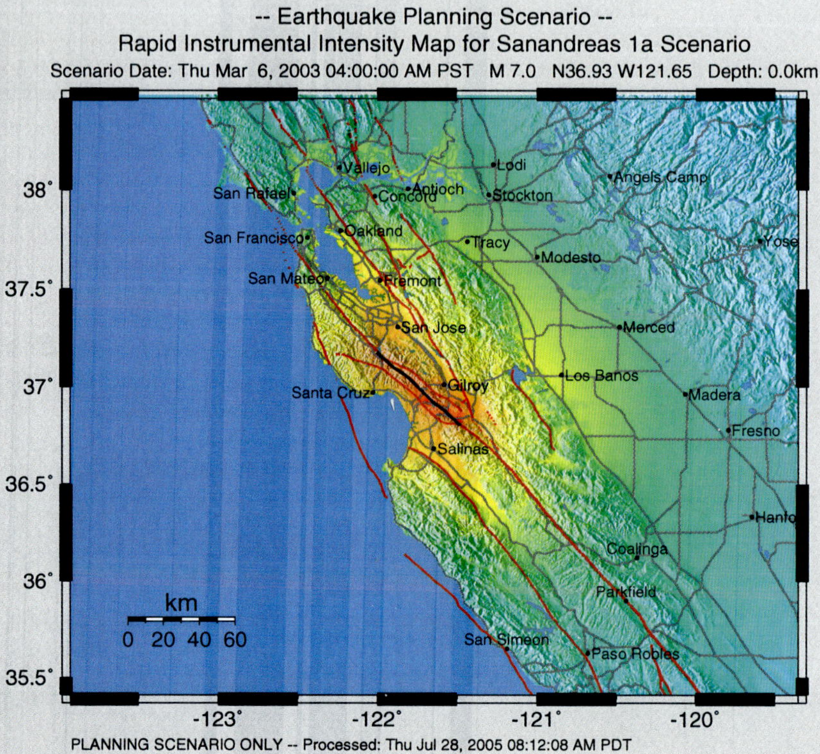

PERCEIVED SHAKING	Not felt	Weak	Light	Moderate	Strong	Very strong	Severe	Violent	Extreme
POTENTIAL DAMAGE	none	none	none	Very light	Light	Moderate	Moderate/Heavy	Heavy	Very Heavy
PEAK ACC.(%g)	<.17	.17-1.4	1.4-3.9	3.9-9.2	9.2-18	18-34	34-65	65-124	>124
PEAK VEL.(cm/s)	<0.1	0.1-1.1	1.1-3.4	3.4-8.1	8.1-16	16-31	31-60	60-116	>116
INSTRUMENTAL INTENSITY	I	II-III	IV	V	VI	VII	VIII	IX	X+

FIGURE 5.2 Scenario ShakeMap for a *Mw* 7.0 earthquake on the San Andreas Fault in the Santa Cruz Mountains.
SOURCE: USGS internet output. See http://quake.usgs.gov/research/strongmotion/effects/shake/SanAndreas_1a_se/intensity.html.

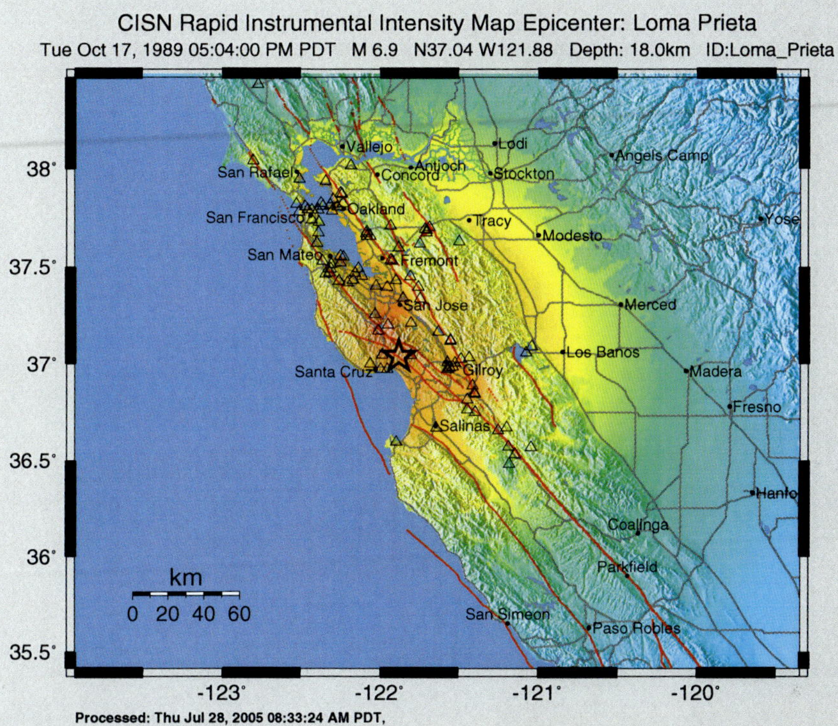

PERCEIVED SHAKING	Not felt	Weak	Light	Moderate	Strong	Very strong	Severe	Violent	Extreme
POTENTIAL DAMAGE	none	none	none	Very light	Light	Moderate	Moderate/Heavy	Heavy	Very Heavy
PEAK ACC.(%g)	<.17	.17-1.4	1.4-3.9	3.9-9.2	9.2-18	18-34	34-65	65-124	>124
PEAK VEL.(cm/s)	<0.1	0.1-1.1	1.1-3.4	3.4-8.1	8.1-16	16-31	31-60	60-116	>116
INSTRUMENTAL INTENSITY	I	II-III	IV	V	VI	VII	VIII	IX	X+

FIGURE 5.3 Actual ShakeMap for the *Mw* 6.9 1989 Loma Prieta earthquake.
SOURCE: USGS internet output. See http://earthquake.usgs.gov/shakemap/nc/shake/Loma_Prieta/intensity.html.

continued

BOX 5.2 Continued

TABLE 5.2 HAZUS-99 Building-Related Direct Economic Loss Estimates for Santa Cruz County

	Building Loss (structural, non-structural, contents and inventory	**Direct Business Interruption Loss** wage, capital-related income, rental, and relocation losses	**Total Building-Related Economic Loss**
Loma Prieta | *Mw* 6.9 ShakeMap	$1,845,100,000	$2,275,000,000	$4,120,100,000
San Andreas Santa Cruz Mountains *Mw* 7.0 Scenario ShakeMap	$789,300,000	$963,600,000	$1,752,900,000
Percentage difference (scenario compared to Loma Prieta)	–57%	–58%	–57%

TABLE 5.3 HAZUS-99 Casualty Estimates for Santa Cruz County

	2:00 a.m.		**2:00 p.m.**		**5:00 p.m.**	
	(most of the population are in residential structures)		(working population is distributed among commercial and industrial structures)		(majority of the working population is in transit)	
	Deaths	**Injuries**	**Deaths**	**Injuries**	**Deaths**	**Injuries**
Loma Prieta *Mw* 6.9 ShakeMap	16	832	91	1,609	34	712
San Andreas Santa Cruz Mountains *Mw* 7.0 Scenario ShakeMap	4	291	25	573	8	243
Percentage difference (scenario compared to Loma Prieta)	–75%	–65%	–73%	–64%	–76%	–66%

earthquake source—the primary sources of uncertainty in the loss estimation process.

In summary, improved seismic monitoring will increase the volume and quality of data available for use in loss estimation models, thereby reducing the uncertainty currently associated with such models. As a consequence, the output of loss estimation models will be more accurate and more acceptable to interested parties (including agency staff, industry managers, homeowners, etc.); be more usable for emergency planning and regulatory purposes; result in better building design, mitigation, and zoning; and result in increases in the amount of pre-event financing and pooling of the earthquake hazard, thereby reducing economic volatility and the crisis nature of post-event recovery.

6

Benefits from Performance-Based Engineering

Earthquake engineering has made significant advances in the past century. What began as an effort to protect lives from future earthquakes has grown to become an effort not only to protect life, but also to minimize damage and functional disruption to levels considered acceptable by owners and the communities in which their buildings are located. Earthquake engineers use "performance-based engineering" procedures to design structures with predictable and defined seismic performance. These procedures have been developed collectively, based on observations of the effects of major earthquakes worldwide. Over the past 50 years, these observations have been aided significantly by the availability of seismic monitoring data—records of earthquake events recorded by weak and strong motion instruments. All components of the built environment—buildings, bridges, roads, utility networks, and dams—share in the benefits of seismic monitoring, even though significant differences exist in how earthquake engineering is approached for each one.

The guidelines, standards, and codes available to earthquake engineers for the design of new structures and the rehabilitation of existing structures hold promise for protecting lives and the built environment against the largest expected earthquakes. Unfortunately, most communities in the United States that are threatened by such earthquakes are doing little to control their seismic risk. This may be partially because such communities—particularly in areas that have not felt shaking in historical time—are often skeptical about the scientific basis of earthquake forecasts and consider that the up-front cost of mitigating the risk is too high (EERI,

2003). The combination of these two circumstances leaves the country in a state of increasing seismic risk. Although the hazard remains the same, the rapid expansion of the built environment nationwide—built without proper regard for earthquake potential—is causing the risk to grow steadily. With the exception of the West Coast, most states with seismic vulnerability do not have adequate building codes or policies requiring seismic design; no state in the nation has an adequate program for mitigating the expected unacceptable performance of existing buildings.

To significantly diminish the growth of seismic risk, radical advances are needed that provide a refined understanding of earthquake hazards, and new analysis and design techniques are needed that more accurately accommodate expected ground motions. Current assessment and design procedures are based on simulation studies, laboratory tests, and post-earthquake field observations that result in generalized and conservative procedures for controlling damage. In recent earthquakes, comparison of building damage with ground motion recordings indicates that buildings have generally performed better than anticipated (Heinz and Poland, 2001). Accordingly, it is reasonable to expect that new techniques can be developed that will reduce seismic design requirements and thereby reduce the cost of seismic safety to more affordable levels. Seismic monitoring records hold the key to understanding how the built environment responds to damaging earthquakes and how best to fine-tune the design process so that the need is adequately—but not excessively—met. Monitoring alone is not sufficient to achieve this goal, but it is certainly a necessary component.

The relatively modest funding required for significantly improving seismic monitoring and the subsequent development of new seismic mitigation techniques should be viewed in light of the potential for reducing the cost of constructing new facilities, strengthening existing structures to achieve proper performance, and avoiding losses after major damaging events. The roughly $200 million investment required for improved seismic monitoring and the cost of continuing research using the records should be viewed in light of the more than $800 billion invested annually in construction, the $17.5 trillion value of the built environment in the United States (FEMA, 2004), and the expected $100 billion plus loss from a single, major earthquake in an urban environment (EERI, 2003).

SEISMIC MONITORING AND THE DEVELOPMENT OF EARTHQUAKE ENGINEERING

Earthquake engineering—the application of science and technology to the design, rehabilitation, and repair of the built environment—has developed over the past hundred years as property owners experienced unacceptable levels of damage caused by ground shaking. The techniques used in earthquake engineering have developed in parallel with the development of seismic monitoring programs. Because the need for design and construction is a daily process, engineers must carry out their work based on their best judgment, adding conservatism to deal with uncertainty. They use material science, structural analysis, large-scale testing, and the observation of failures to extend their understanding and fuel their intuition, and it is their collective intuition that is the basis for the prescriptive building standards and codes in use today.

The goals of earthquake engineering have also evolved over the past hundred years. The earliest earthquake engineering attempts were undertaken both to eliminate damage and to neutralize the concerns raised by developers, bankers, and insurance companies. After the 1906 San Francisco, the 1925 Santa Barbara, and the 1933 Long Beach earthquakes, mandatory building codes emerged as a tool for eliminating the damaging effects of ground shaking. Seismic monitoring focusing on structures began in earnest during the late 1920s, and has grown steadily ever since. After the 1952 Kern County earthquake, a more rational goal emerged that was oriented toward protecting lives and keeping damage to repairable levels for all but the largest earthquakes. This goal remains the underlying principle of earthquake engineering today, and the goal of protecting lives has taken on a very broad definition.

Seismic monitoring programs that are designed to provide improved understanding of the actual effects of earthquakes on the built environment are less than 50 years old. This, of course, is only an instant in geologic time, and so far only a small number of useful data sets have been obtained worldwide. Although much has been learned from those data sets, provoking dramatic changes to the design process, few records exist for sites that have experienced catastrophic damage from a design-level[1] earthquake. The most important information that seismic monitoring networks are expected to provide has yet to be recorded.

Seismic monitoring has led to development of national seismic hazard maps that are used to design structures with appropriate strength and durability. Although engineers once suggested—after the 1906 San Francisco earthquake—that the entire nation shared a common threat from earthquakes, it is now understood that the seismic hazard varies dramati-

---

[1]A design-level earthquake represents the strongest ground motion expected to occur with a specified exceedence probability during the life of a building. Recent earthquakes in Kobe, Japan, and Taiwan were design-level earthquakes that provided a variety of strong motion records that have been useful for understanding the performance of structures in the immediate vicinity of the instrument.

cally across the nation. For example, whereas all of California and much of Nevada were once considered to be areas of highest hazard (ICBO, 1973), a combination of monitoring and scientifically defensible risk assessment has permitted the hazard to be mapped according to multiple zones that vary from moderate to highest hazard. Only the thin zone paralleling California's great faults is now believed to be an area of highest hazard.

Strong motion recordings have allowed seismic design procedures to be based on measurable parameters. Spectral values have replaced peak ground acceleration estimates as the key indicator of the severity of the earthquake hazard. Before monitoring systems were designed and installed, engineers could only speculate about the intensity of shaking based on the resulting damage. Now, in-structure instrumentation yields records that permit a clearer understanding of structural response during strong shaking, much as the "black-box" in an aircraft provides key information about conditions just prior to a crash. Monitoring provides insights into the characteristics and strength of the shaking that has caused damage. The few available seismic monitoring records show that structures experience earthquakes in a highly dynamic and time dependent manner, with the resulting structural performance due to complex combinations of the loading history, material strength, and structural design.

Earthquake engineering design techniques have improved after each damaging earthquake and resulted in increasingly more advanced seismic design standards. When an earthquake occurs and structures experience more damage than their owners and engineers judge acceptable, the community of engineers adjusts design standards to avoid a repeat occurrence. Unfortunately, these advances in design are limited by the quality and quantity of the seismic monitoring records collected. When there are no records, there is a tendency to apply the new techniques to all buildings in all seismic environments. When there are records, the changes often apply only to construction in areas that are expected to experience a similar level of shaking. For example, earthquake engineers developed new procedures for the design and construction of moment-resisting steel-frame buildings based on the unexpected damaged that occurred during the 1994 Northridge earthquake. Unfortunately, no records were available at the sites where significant damage to moment-resisting steel-frame buildings occurred, so the subsequent research and resulting recommendations had to be based entirely on estimates of ground shaking. The resulting recommendations apply to the design of steel buildings nationwide. While they represent an important step forward in the design process, it is likely that they are more conservative than needed, mainly because strong motion records were not available to calibrate the observed damage.

The 1971 San Fernando earthquake produced the first regional set of strong motion records related to structures. The combination of seismic monitoring records and observations of damage resulted in code provisions requiring stronger, more resilient buildings that would experience less damage; the design of "essential buildings" that would remain functional; and the development of techniques for dealing with the massive inventory of old buildings that were constructed without sufficient seismic resilience. The San Fernando earthquake also provided the first evidence that structures that were designed using the same techniques did not necessarily perform in the same manner. It appeared that there were major differences in the ground motion from block to block, and significant differences in building performance due to the materials used in construction, the size and shape of the buildings, and the quality of construction.

In order to adequately initiate the development of the next generation of design procedures to achieve the goals of earthquake engineering, specific recorded information is needed for a variety of construction styles and geologic conditions. After the 1971 San Fernando earthquake, seismic monitoring programs at federal and state levels were expanded to begin to provide the needed information. By the time of the 1989 Loma Prieta and 1994 Northridge earthquakes, hundreds of instruments had been deployed in the western United States and hundreds of records were collected. These instruments were focused on collecting three fundamentally different data types—free-field ground motion, lifeline response, and building response. They were deployed primarily in the most seismically active regions (i.e., the West Coast), with the expectation that the results could be extrapolated to all forms of construction nationwide.

In parallel, earthquake engineers developed a variety of approaches to seismic design and rehabilitation that included permitting old, nonconforming buildings to remain in use or, in some cases, to be rehabilitated to a minimum "life safety" level. New buildings with normal use occupancy were designed to be "life safe" and "repairable" in the event of a large earthquake, and essential buildings were designed to be capable of operating after a major earthquake. While these standards were most often applied in California, they were also applied—especially for federally owned buildings—in other seismic regions. These various approaches to earthquake engineering were the forerunners of performance-based engineering.

IMPROVEMENTS IN SEISMIC MONITORING NEEDED TO SUPPORT PERFORMANCE-BASED ENGINEERING

Performance-based engineering began its formal conceptualization for practicing engineers after the 1994 Northridge earthquake, when the

Structural Engineers Association of California (SEAOC), with a grant from the Federal Emergency Management Agency (FEMA), produced *Vision 2000: Performance Based Seismic Engineering of Buildings: Interim Recommendations* (SEAOC, 1995). These recommendations were developed in response to the concern that the $40 billion cost of the Northridge earthquake was too high and the damage too disruptive to local communities. It appeared that the traditional goals of earthquake engineering related to protecting life were not sufficient. At the heart of the three volumes of data and interim recommendations in SEAOC's report is the call for design procedures that produce buildings capable of performing at any one of a variety of predictable levels. The process has made the expected performance of structures during earthquakes more visible and has provoked discussion about the requirements for acceptable performance.

The instruments deployed prior to the Loma Prieta and Northridge earthquakes yielded numerous strong motion records, resulting in opportunities to begin correlating observed damage with strong motion recordings and structural analysis techniques. Unfortunately, in both earthquakes there were no records taken in the immediate vicinity of any sites of major damage, and there were no records from significantly damaged bridges or other critical facilities.

Detailed surveys conducted after the 1994 Northridge earthquake in the immediate areas surrounding the strong motion instruments illustrated once again the widely varying degrees of damage: much more variation was observed than would be expected based on the design standards in use (ATC, 2001). Unfortunately, the expanded instrumentation programs developed based on the 1971 San Fernando experience again failed to yield sufficient information to develop a credible understanding of the variations in ground motion and the reasons buildings performed as they did. The vast majority of buildings performed better than expected, so any notion that these records formed a basis for nationwide analysis and design was lost. Had there been thousands of instruments and records in the areas of greatest damage, there is no doubt that additional new advances would have been made in earthquake engineering with the consequent potential for considerable economic benefits.

Another example of the inadequacy of the current seismic monitoring programs resulted from the 2003 San Simeon earthquake. Because this was an area of only moderate seismicity, few ground motion instruments and a single building instrumentation package had been installed. Given the intensity of the shaking and the lack of damage, an excellent opportunity to gain additional understanding of the damaging potential of moderate earthquakes was lost—an understanding that could have been applied in other areas of moderate seismicity throughout the nation. Opportunities for advances in seismic mitigation programs and techniques will continue to be lost until sufficient instruments are deployed nationwide to capture the key characteristics of every damaging event.

The development of performance-based engineering has resulted in a discussion concerning the level of damage that is expected, and provided the opportunity for owners and communities to determine what level of damage they will accept. Unfortunately, the engineering techniques needed to design to specific performance levels are still being developed. Efficient and cost-effective design standards are needed, because most procedures in use today yield expensive solutions that serve as deterrents to action.

Needed Improvements

To achieve the benefits that improved seismic monitoring can potentially provide, the following six enhancements to the current monitoring program are needed:

1. The addition of sufficient free-field instruments nationwide to develop an understanding of the relationship between the source characteristics of an earthquake and the strong motions that are produced. Instruments also have to be located to identify the effects of the geologic setting and the local site conditions on ground motions, with a special focus on the seismic zonation of urban areas to identify locations where unusually strong ground motions are expected to occur. The end result that is required is a set of characteristic waveforms that can be used in design, based on an appropriate probabilistic assessment. These will provide the best possible input for an efficient design or assessment process and permit the proof testing of buildings in regions that experience the design-level ground motion.
2. The addition of sufficient monitoring to identify the ground motion characteristics that trigger liquefaction, lateral spreading, and landslides. This will also allow the risk to be calculated in a manner consistent with similar calculations for the design of structures. This is important because the mitigation of geologic hazards is often carried out from a deterministic perspective, without regard for the probability of occurrence. Substantial savings in foundation costs, as well as expanded opportunities for building on otherwise questionable sites, will result.
3. The addition of sufficient urban monitoring (i.e., the ideal extension to the present Advanced National Seismic System [ANSS] proposal) so that every structure that is damaged in an earthquake has a reference record that is suitable for understanding its performance. This does not require an instrument in every building—rather, there should be an instrument sufficiently close to record the ground motion that was experi-

enced. This equates to at least one instrument in every zip code and one instrument on every active geologic structure (fault or fold) within that zip code. The resulting records would provide the minimum amount of consistent information needed to understand the earthquake motion and provide the opportunity to develop the statistical data necessary both to calibrate assessment techniques and to develop appropriate performance indicators. The urban monitoring instruments proposed for the ANSS will partially accomplish this goal.

4. The addition of sufficient structural monitoring of enough buildings nationwide to fully document the performance of all common building types during an event in terms of the lateral forces resisted, displacements experienced, the location and demand on elements developing ductility, and foundation-soil interaction. Instrumentation must allow for a complete determination of the demand on all structural and nonstructural elements. The resulting records will, in time, provide the data needed for the development of new analysis techniques that fully capture the linear and nonlinear performance of the structure. Current techniques are unable to estimate the deterioration of structural elements under strong shaking and therefore often overstate the significance of damage. In order to minimize the cost of seismic design and rehabilitation, more accurate techniques for estimating damage are needed. Special emphasis should be placed on instrumenting publicly owned buildings, especially federal buildings, to ensure continuity of maintenance of the instruments, open access to information about structural design and construction history, timely access to the monitoring records and to the buildings themselves so that recorded shaking levels can be correlated to actual building damage, and avoidance of liability issues that may concern private building owners.

5. The development and deployment of new methods for monitoring buildings to directly record inter-story drift[2] demand at critical locations from both structural and nonstructural perspectives. Current building instrumentation packages record acceleration, and integrate the waveforms to determine velocity and displacement. There is considerable controversy surrounding the accuracy of the calculated displacements, especially when they are used to calculate inter-story drift. Directly measured inter-story drift is expected to provide the most reliable ability to assess damage potential.

[2]Inter-story drift is the amount of horizontal movement that occurs between floors during earthquake shaking. For example, if the tenth floor of a building deflects 20 inches and the ninth floor deflects 18 inches at the same time, the inter-story drift between the ninth and tenth floors is 2 inches.

6. Lifeline systems include transportation, water, wastewater, electric power, telecommunications, and gas and liquid fuel systems. They must perform successfully as complete systems to ensure uninterrupted operation of essential services. The addition of sufficient monitoring of lifeline systems to fully capture the interdependence of the related structures (e.g., pumping plants) and the interconnecting components (e.g., piping) in their distributed environment is required. This will allow a full understanding of the source and impact of element failures in the system that will lead to more robust designs.

CALCULATION OF BENEFITS PROVIDED BY PERFORMANCE-BASED ENGINEERING

When assessing the value of improved seismic monitoring as it relates to performance-based engineering, three parameters must be considered. These include the value of the built environment within the United States, the rate of construction, and the annual expected loss from earthquakes. Current estimates suggest that there are $17.5 trillion worth of structures and that 80 percent of construction is residential.[3] The value of structures in states within high and very high seismic zones is about $5.8 trillion (33 percent of the total) and, when all states prone to seismic damage are included this amount, increases to about $8.6 trillion (49 percent of the total). For the purposes of this study, it is reasonable to assume that the average value of an instrumented building is $5 million. Construction estimates for 2004 totaled about $500 billion dollars for buildings and $400 billion for lifelines, and the annual construction value is expected to exceed $1 trillion per year within the next 10 years (FMI Corporation, 2004). As noted above, total annualized earthquake losses throughout the United States are estimated to be about $5.6 billion per year for buildings and building-related costs (FEMA, 2001a), and a single, major urban earthquake is expected to cause losses of more than $100 billion (EERI, 2003).

Other critical assumptions about the built environment relate to the cost of seismic design and the cost of seismic rehabilitation. In an effort to establish a potential "ballpark" estimate of the benefits of improved monitoring, it is useful to consider the cost of seismic mitigation in general. Only broad-brush estimates are needed to encompass the various design styles and performance-based engineering techniques. Recent experience in various design practices has suggested (anecdotally) that the cost of including seismic design can range from 1 to 10 percent of a project

[3]Based on the national inventory and valuation models contained in HAZUS-MH, released in early 2004.

budget. For example, a building that costs $150 per square foot to build without seismic design features would cost an additional $2 to $15 per square foot (a total of $152-165 per square foot) with seismic design. The difference relates mostly to the performance level selected, the sophistication of the design team, and its willingness to incorporate seismic design in the conceptual framework of the project and work to minimize its impact. Similarly, design office experience suggests that the cost of seismic rehabilitation can range from 10 to 150 percent of the replacement cost of the structure, depending on the structure, its condition, the seismic performance objective selected, and whether the structure will be occupied during reconstruction. A good generalized average cost is 20 percent of the replacement value.

Another important assumption relates to the number of existing buildings that need seismic strengthening. Of the 50 states and the District of Columbia, 42 have some degree of earthquake potential, and 18 are considered to have high or very high seismicity. Based on a variety of building inventories and extensive seismic rehabilitation experience, it is reasonable to assume that about 10 percent of existing buildings within the earthquake-prone areas of the United States need seismic strengthening. Finally, since the majority of buildings remain in use until destroyed by natural disasters or neglect, all cost savings were calculated under the assumption that all buildings would eventually experience a design-level earthquake. The total value was then translated to an annualized cost by multiplying the total cost by 0.04.[4]

Seismic monitoring programs in place today will continue to generate benefits from performance-based engineering to the extent that they capture and record damaging events. The proposed improvements to the monitoring program are considered in terms of incremental improvements in seismic monitoring capabilities. The first is the augmentation of the United States National Seismic Network (USNSN) seismic monitoring backbone that will be provided as a component of USArray. The second is the implementation of the initial phase of the ANSS program, and the final step would be to add sufficient seismic monitoring nationwide to ensure that every damaging earthquake that occurred would be recorded to the extent necessary to advance the engineering design standards as much as possible.

If these proposed enhancements are not done, the existing networks will continue to deteriorate due to age and obsolete technology, and eventually little seismic monitoring will exist to capture data from future

---

[4]An approximate annualized value is derived by multiplying the dollar value of the capital stock by a plausible long-term interest rate, estimated as 4 percent.

earthquakes. The current implementation of USArray should lead to a long-term improvement in the understanding of the seismic hazard nationwide, although it will not provide any additional information related to the performance of structures in damaging earthquakes or provide any immediate benefit to the hazard assessment of the nation used in structural design. The implementation of ANSS—required to maintain what is currently available and achieve some of the enhancements stated above—should generate at least six significant benefits discussed below: two that are short-term (1 and 2) and four that are intermediate- to long-term (3 to 6). The extent and timelines of achieving these benefits will depend on significant earthquakes occurring in areas that are instrumented, as well as the timing of program funding.

1. *Proof testing of instrumented buildings:* Buildings that are instrumented and experience damaging earthquakes will provide new insights into how to better design buildings to predictable performance levels. They also will be candidates for "proof testing," in that their performance capability for the recorded event will serve as a benchmark for performance during other earthquakes.

The December 22, 2003, San Simeon earthquake may have been close to the maximum likely earthquake for that area and provides an example of this proof-testing benefit. The U.S. Geological Survey (USGS) probabilistic seismic hazard maps define design-level events for that part of the central California coastal area in terms of peak ground acceleration, as well as for short-period and 1.0-second spectral accelerations. Templeton Hospital, located in the area of strongest shaking, was instrumented and recorded strong motions at about the maximum design level expected. The building experienced only slight damage and did not experience any disruption of function. Like all hospitals in California, this building is currently slated for strengthening to meet new and stringent state requirements at an estimated cost of $20 million. Because the building has been essentially proof-tested, and the records of this testing are available, it is likely that no seismic strengthening is needed. Accordingly, about $50,000 worth of instrumentation and about $50,000 worth of instrument maintenance over the past 20 years will likely yield a 200-times benefit.

Currently, approximately 300 buildings are instrumented nationwide, and this number will increase to approximately 600 under the ANSS program.[5] As many as 50 percent of the instrumented buildings that need

[5]The original calculations in a prerelease draft of this section were based on instrumentation of 3,000 buildings nationwide by ANSS. Clarification of implementation plans provided by USGS indicates that approximately 300 buildings will be instrumented with multiple sensors, so this figure is used in the calculations that follow.

rehabilitation are expected to pass their proof test once a design-level earthquake occurs. Engineering experience indicates that this will translate into an annualized savings of $3 million under ANSS and $665 million if all critical and economically significant buildings at risk of earthquake damage are eventually instrumented.

2. *Post-earthquake repair of instrumented buildings:* Structural engineers responsible for evaluating the post-earthquake condition of a building that is instrumented will have the advantage of knowing what level of ground shaking caused the observed damage, and they can determine how the shaking compares to the event for which the building was designed as well as the event that it has to be repaired to resist. This information will generally lead to lower repair costs, because the adequacy of the existing building will be better understood along with its key vulnerabilities—the repair and rehabilitation efforts can be better focused to address actual deficiencies. Engineering experience indicates that repair cost savings—ranging from 5 to 20 percent—are expected to occur for 20 percent of the currently instrumented buildings and 30 percent of the buildings to be instrumented under ANSS. Based on building inventory estimates, this is expected to translate into an annualized saving of $2 million.

3. *Improved seismic hazard maps:* The decision to design for seismic conditions—or rehabilitate because of seismic conditions—depends first on an understanding of the hazards anticipated at a particular building site. Although detailed site-specific seismic hazard studies can be performed, the costs of such studies are too high for most building projects. Seismic hazard maps have been available for decades to allow for a less rigorous assessment of seismic risk. Scientifically defensible maps were produced by USGS in 1997 that—for the first time—used seismic monitoring data in conjunction with engineering-based parameters. The limited distribution of the seismic monitoring data on which these maps are based has meant that they can only be used accompanied by a number of assumptions that lead to conservative assessments. Even with these limitations, these maps have refined our understanding of the nationwide distribution of earthquake hazards. Improved seismic monitoring using adequate free-field instruments is a critical requirement for further refining these maps. In addition, there is a need to better understand the relationship between the particular source characteristics of an earthquake and the strong ground motions that are produced. Instruments have to be located to identify the effects of the geologic setting and local site conditions on ground motions, with a special focus on the seismic zonation of urban areas to identify locations at which unusual ground motions may occur. Improved seismic monitoring of weak and strong ground shaking will ultimately lead to improvements in the hazard maps used for design. The expected improvements will have a direct impact on the cost of construction and the level of

damage experienced in those states that include areas of high or very high seismicity. Engineering experience suggests that an additional 1 percent savings in construction cost could occur with implementation of each of the incremental seismic monitoring programs—USArray, a revitalization of the USNSN, and full implementation of ANSS—as they provide improved seismic hazard information. This would translate to an annualized savings of $49 million for each of the incremental programs.

4. *Refined analysis techniques:* The ability of engineers to predict the performance of buildings during earthquakes depends on their ability to model the behavior for a specific ground motion. The techniques currently in use are based on available mathematical formulations and material properties. For the most part, they do not benefit from the combination of damage observations and recorded motion. There is evidence of considerable uncertainty in current predictions—using the best available analysis techniques—of the way buildings experience damage. Improved monitoring of buildings can eliminate this uncertainty, once sufficient waveforms are recorded and analysis techniques developed. This will ultimately result in reduced construction costs for new designs and for the rehabilitation of existing buildings. To achieve this goal, it will be necessary for sufficient urban monitoring to be added so that every structure that is damaged in an earthquake has a reference record that is suitable for understanding its performance.

In addition, sufficient structural monitoring has to be added to enough buildings nationwide to fully document the performance of all common building types in terms of the lateral forces resisted, displacements experienced, the location and demand on elements developing ductility, and foundation soil interaction. Instrumentation must allow for a complete determination of the demand on all structural and nonstructural elements. Neither the current programs nor the enhancements to be provided by USArray and the revitalized USNSN will add this monitoring—it will only occur after full implementation of the ANSS program.

The most significant impact of refined analysis techniques will be felt in rehabilitation projects. Based on an expectation that the cost of rehabilitation will decrease on the average by 3 percent, and considering the 10 percent of the inventory that is expected to need strengthening, this translates to an annualized savings of $34 million dollars.

5. *Improved procedures for new construction:* Improved seismic monitoring will provide the records needed to calibrate the earthquake engineering process, remove conservatism as appropriate, and reduce the cost of construction. This portion of the potential savings is interconnected with the ongoing research and testing programs related to seismic mitigation. The benefit is expected to manifest itself as a reduction in the cost of seismic mitigation and a reduction in the loss expected, with the latter anticipated

to be most significant. It is reasonable to expect that if the ANSS program is fully implemented, the average cost for new construction will decrease by 1 percent for 30 percent of the buildings built in the very high and high seismic regions. This translates into an annualized saving of $20 million.

6. *Improved procedures for rehabilitation of existing construction:* From the decision to consider seismic hazards, to the process for identifying the seismic deficiencies in a building, to the actual techniques for rehabilitating buildings, improved seismic monitoring will enhance understanding, reduce conservatism as appropriate, and make seismic mitigation more affordable and acceptable. The current assessment techniques, when judged against the performance of buildings in large earthquakes, appear to substantially overpredict the expected damage when seismic monitoring information is available to quantify the intensity of shaking. This tendency toward overprediction, when applied to an inventory of existing buildings being considered for seismic rehabilitation, will also overpredict the number that have to be strengthened. Unfortunately, this often results in very high estimates for the cost of mitigation that result in no action. The same thing often happens when communities attempt to develop public policies aimed at mitigating their seismic risks.

The Unreinforced Masonry Building Ordinance adopted by the City of Paso Robles, when taken in the context of how buildings performed in the 2003 San Simeon earthquake, provides a good example. As required by California law, the City of Paso Robles had inventoried its unreinforced masonry buildings (UMB) and set a time schedule for their rehabilitation that extended through 2017. Unfortunately, the earthquake occurred sooner, and one building collapse resulted in two deaths. Of the approximately 20,000 buildings in and around Paso Robles, all but a handful performed without significant damage, even though they probably experienced ground motions near or above the design level. Current procedures for evaluating the seismic strength of buildings in that region, however, would probably show that a number of these buildings were not strong enough and were candidates for retrofit. Seismic monitoring will lead to improved assessment that will, in turn, lead to an increased focus on the buildings that actually have to be rehabilitated, thereby bringing rehabilitation program estimates down to a more acceptable size and allowing for inventories of hazardous buildings to be scheduled for mitigation, prioritized by their risk. In the case of Paso Robles, it appears that the one collapsed building should have been designated for rehabilitation many years ago. It is reasonable to expect that the annualized savings related to improved rehabilitation techniques will be similar to that calculated for refined analysis—an additional $34 million once the ANSS program is in place.

TABLE 6.1 Summary of Potential Design and Construction Benefits from Improved Seismic Monitoring

Benefit	Buildings Affected	Total Value	Seismic Cost[a]	Rehabilitation Cost Saved	Annual Savings	Beneficiary
Proof testing of instrumented buildings[a]	300 added by ANSS	$3 billion	$150 million	$75 million	$3 million	Building owner
Post earthquake repair[b]	300 added by ANSS	$3 billion	$315 million	$63 million	$2 million	Building owner, FEMA
Improved seismic hazard maps[c]	All buildings in seismic zones	$165 billion	$4.9 billion		$49 million	Building owner
Refined analysis techniques[d]	10% of existing inventory	Annual $170 billion	$34 billion	$850 million	$34 million	Building owner, FEMA
Improved new construction procedures[e]	All buildings in seismic zones	$165 billion			$20 million	Building owner, FEMA
Improved rehabilitation procedures[f]	10% of existing inventory	Annual $170 billion	$34 billion	$850 million	$34 million	Building owner, FEMA
Total annualized savings					$142 million	

[a] Seismic cost is the cost to add appropriate seismic strengthening to a building during repair, rehabilitation, or initial construction.
[b] 50 % proof-tested, saving is from eliminating the need to rehabilitate.
[c] 20 % less repair costs.
[d] 1 % reduction in seismic cost.
[e] 5 % reduction in seismic cost.
[f] 2 % reduction in seismic loss for 30 % of the buildings.

SUMMARY

This discussion has been rooted in observation and experience in the design and construction environment. It has attempted to generalize the issues sufficiently to allow simple estimates to be made that subjectively quantify the value of improved seismic monitoring. The total annualized savings described above amount to more than $140 million per year after implementation of the ANSS (summarized, together with identification of the benefit recipients, in Table 6.1). These calculations were also performed for the situation where 3,000 buildings nationwide are instrumented, rather than the approximately 600 planned as part of the existing ANSS proposal. These calculations indicate that such an expanded instrumentation program would provide potential annualized savings of about $250 million.

7

Benefits for Emergency Response and Recovery

Seismic monitoring—and particularly the products derived from monitoring—provides important input for at least three basic components of emergency response: (1) response readiness, or the capacity of an organization to respond effectively to large earthquakes as reflected in planning and exercises; (2) management of earthquake emergencies, or the ability of an organization to mount a timely and effective response that minimizes the loss of life and property damage, and maintains operational capabilities; and (3) rapid recovery, encompassing the mitigation of hazards, restoration of the built environment, and return to normal community life. This section describes the impact of seismic monitoring on these three elements of emergency response and recovery, and examines the potential benefits of enhanced monitoring.

It is important to keep in mind that both the benefits of seismic monitoring and the examples that are cited in this section currently apply only to the limited region—essentially urban areas of California—where state-of-the-art networks are currently in operation and the density of network coverage is sufficient to provide relatively accurate information for emergency response and recovery. Other regions, despite their identification as areas of high or moderate seismic potential, currently realize considerably fewer benefits due to the lack of modern digital seismic and strong motion instrumentation, the lack of adequate station coverage to produce response-relevant products, or both.

MONITORING FOR RESPONSE READINESS

Organizations and individuals in seismically vulnerable areas of the nation seek to reduce risks in diverse ways, including informed land-use planning, structural and nonstructural mitigation, the purchase of insurance, and planning for response and recovery. It is this last strategy for risk reduction that is the topic of this section. Response readiness—action associated with development of plans and conducting exercises—relies on seismic monitoring to describe long-, intermediate-, and short-term earthquake potential, to provide information on earthquake effects, and to estimate the impacts of earthquakes on the community.

For areas of high seismic risk, the U.S. Geological Survey (USGS), state-level geological agencies, and academic scientists have used monitored information to develop projections of long-term seismic potential based on fault characteristics, local geological conditions, recurrence intervals for large events, and other factors (WGCEP, 1988, 1995, 2003). These studies have provided state and local government agencies and private sector entities with essential information for earthquake hazard reduction actions that include focused planning and prioritized hazard mitigation for areas judged to have the highest probability of large damaging earthquakes. These studies have also provided the basic information needed for the development of exercise scenarios to improve response readiness.

Organizations that must respond rapidly to a significant earthquake or other disasters typically conduct drills and exercises to test their readiness and the quality of their planning. These exercises vary in extent of involvement and degree of detail, from limited "tabletop exercises" that include a few key decision-makers gathered around a conference table, to full-scale field exercises that include many departments and full activation of emergency operations centers. With the development of ShakeMap, emergency response exercise scenarios have reached degrees of sophistication that provide substantial benefits for response readiness.

When combined with loss estimates from HAZUS, ShakeMap earthquake scenarios provide important details of potential earthquake impacts on a city, county, or region. In the past, response organizations relied on vague projections that were largely the result of guesswork regarding earthquake size and effects. The currently available ShakeMap-HAZUS scenarios include empirically grounded estimates of shaking intensity and regional patterns of shaking, and as input data for HAZUS, ShakeMap contributes to plausible estimates of total dollar losses, utility damage, building damage, and population impacts including the number of deaths, various levels of injury, the number of people displaced from their homes, and the probable demand for shelter and mass care. When mapped, these estimated losses supply exercise participants with details that provide far more rigorous tests of response procedures, protocols, and plans than has been possible in the past. It is reasonable to assume that better scenarios and thus better exercises will result in greater response readiness.

Monitored seismic information also contributes to response readiness in short- and intermediate-term time frames. Although earthquake prediction has not developed as rapidly as was once hoped (see Chapter 4), monitored seismic data are crucial for identifying "hot spots" and seismic activity that could be precursors to large damaging events. Historically, such activity has triggered specific emergency response preparations. In California, there are reasonably well-defined areas in which earthquake activity or other monitored changes could greatly increase the short-term probability of large damaging earthquakes. One such hot spot is the southern terminus of the San Andreas Fault along the eastern shore of the Salton Sea. Earthquake swarms in this area or a single moderate-size event ($Mw \geq 5$)—if interpreted by scientists as potentially precursory to a large event on the San Andreas Fault—would trigger activation of plans that require specific actions by the California Governor's Office of Emergency Services (OES) and other agencies. Based on short-term changes in seismic activity, the USGS has recently commenced real-time forecasts, updated every hour, of earthquake hazard in California for the following 24-hour period.[1]

The State of California has developed a short-term earthquake advisory plan (OES, 1990) that contains detailed procedures for emergency activation and public warning based on scientific interpretation of monitored data. This plan calls on the OES to notify state agencies and local jurisdictions in the area of enhanced seismic risk and to issue a press release to the print and electronic media recommending actions designed to enhance readiness to respond should damaging earthquakes occur. The plan identifies methods for transmitting the advisory message and includes templates and sample messages to facilitate issuance of advisories in a timely manner, given that the advisory period will be from 3-5 days.

REAL-TIME INFORMATION FOR EMERGENCY RESPONSE OPERATIONS

The emergency response phase of an earthquake spans a period of time ranging from the initiation of the earthquake to the conclusion of activities designed to save lives, treat the injured, shelter the displaced, and assess the damage. Monitored information from seismic networks is

---

[1]See http://pasadena.wr.usgs.gov/step.

of critical importance during this phase, particularly in the initial stages. Beginning with indications of the occurrence of an earthquake fault rupture and continuing into the first few hours of the response, monitored data have a potentially wide-ranging impact on the timeliness, efficiency, and efficacy of emergency response.

Several nations—including Mexico, Taiwan, and Japan—have employed monitoring networks to alert user communities that potentially damaging ground motion is approaching from a distant earthquake. These "early-warning systems" may provide from a few seconds to a few tens of seconds warning, thus facilitating life safety and rapid hazard mitigation actions. Both the network operators, and those who have studied behavioral responses to earthquake early warning systems (e.g., Tierney, 2000; Shoaf and Bourque, 2001) have identified a number of actions that can be taken in response to a warning. These actions include taking cover, moving away from hazards, evacuating vulnerable buildings, programming elevators to stop at the next floor and open doors, shutting down computer systems, slowing down or stopping trains, and a range of other possible activities. Although earthquake early-warning systems based on modern digital seismic networks could be implemented in the United States and are being planned as part of future Advanced National Seismic System (ANSS) activity (Leith, 2005), such systems require the dense seismic networks and real-time telemetry that currently exist only in California's two major urban areas.

Although early earthquake warnings are not yet a reality in the United States, warning systems for tsunamis have been in place since 1948. As noted in Chapter 4, the National Oceanic and Atmospheric Administration's (NOAA) National Weather Service operates two tsunami warning centers, one in Palmer, Alaska, and the other in Ewa Beach, Hawaii. The goal of these centers is to rapidly and accurately identify events that could trigger a tsunami and provide timely warnings to coastal communities that may be impacted. These warnings, if accurate and timely, can save lives based on rapid evacuation of areas likely to be affected by damaging tsunami waves.

It is in the immediate post-impact period that information from seismic networks has the greatest potential benefit for emergency response. In areas where dense arrays of modern digital instruments are deployed, emergency response agencies now have resources available to them that will affect all aspects of emergency response—resources that were not available during the Northridge earthquake of January 17, 1994, an event that remains one of the costliest natural disasters in the nation's history. After the Northridge earthquake, monitored seismic information, consisting of magnitude and location, played a relatively minor role in response. The significant enhancement of the Southern California Seismic Network under the 1997-2001 TriNet Project resulted in improvements that greatly increased the quality and quantity of information available to responders in the minutes and hours following a future damaging earthquake.

The very rapid availability of earthquake source data—including magnitude, location, depth, and fault geometry—provides basic orienting information for emergency responders, essential information for the news media and the public, and input data for other applications and response-relevant products. Maps of ground shaking intensity (ShakeMap) have many important applications in emergency management. Because ShakeMap is available via the Internet, all emergency responders at all levels of government and the private sector have access to the same rapidly available information. With this information, responders can quickly assess the scope of the emergency and mobilize resources accordingly. Early reconnaissance efforts can target areas known to have been shaken most severely, and key emergency services including search and rescue, emergency medical response, safety assessment of critical facilities, and shelter and mass care can be expedited based on a more rapid identification of incident location. Monitored information is also useful for rapidly assessing situations in which a large, widely felt earthquake occurs but causes little damage (such as the Hector Mine earthquake of October 16, 1999). Clearly, there are significant economic benefits in scaling a response to the consequences of an event, including no response for an earthquake that requires none.

Based on ground motion data from seismic networks, HAZUS can be used to generate estimates of economic losses, utility system damage, and population impacts. These data provide information useful for several response and recovery actions (discussed in the following section). For response actions, HAZUS outputs will supplement the ground shaking information from ShakeMap by estimating the level of damage to buildings, utilities, and transportation infrastructure; contribute to identifying the appropriate response activation level; provide guidance for initiating building safety assessments; help determine the number of shelters that must be opened to house the displaced; and provide an estimate of the amount of resources needed to care for people in temporary housing.

Monitored data from instruments on or near buildings and other structures can also provide the means to conduct real-time damage assessment of critical infrastructure (e.g., hospitals, highway bridges, emergency operations centers) and avoid secondary hazards (e.g., from dams, natural gas or petroleum pipelines, etc.). Emergency response based on monitored data can be used for rapid or automated closure of damaged bridges, evacuation of areas vulnerable to dam collapse, actions by utility operators to preserve electric power and gas, and rapid notification to emergency medical transportation units regarding the damage status of hospitals and trauma care centers. These data might also be used to prioritize safety and damage assessment, repair, and restoration.

Response operations—including search and rescue, shoring or demolition of damaged structures, and safety inspections—are vulnerable to secondary hazards posed by aftershocks and additional earthquakes that are triggered by the initial ground shaking. Rapidly available information from seismic networks can reduce exposure to these risks. The location and magnitude of large aftershocks and triggered events can be estimated using models (Jones and Reasenberg, 1989; Reasenberg and Jones, 1994) based on the observation and analysis of an evolving sequence of earthquakes. This information can be used to establish cordons, restrict access, and warn responders. After the 1989 Loma Prieta earthquake, portable instruments were placed near the epicenter in the Santa Cruz Mountains and alerts were broadcast in real-time to crews working on the collapsed portions of the Nimitz Freeway (Interstate 880), providing 10-20 seconds warning that ground motion from aftershocks was imminent (Bakun et al., 1994). Aftershock location and magnitude estimates, in combination with alerts based on monitored information, can provide an additional margin of safety and risk reduction for responders.

Reliable operation of seismic monitoring networks is a critical requirement for providing earthquake information to the emergency management community. Like all lifeline infrastructure, seismic networks are dependent on electric and telecommunication utilities as well as the Internet. Consequently, efforts to upgrade the national monitoring capability should include long-term maintenance of existing instruments and communications infrastructure as well as deployment of new instruments.

MONITORING FOR EARTHQUAKE RECOVERY

A popular saying among emergency managers regarding earthquake disasters is that "recovery begins when the shaking has ceased." Planning for recovery must begin immediately and concurrently with response activities. Although response is fundamentally local in character, recovery from a large damaging earthquake is regional and national in scope, involving every level of government as well as the private sector. Recovery decisions involve budgetary allocations and the activation of programs for individual victim support, long-term housing reconstruction, business and economic recovery, and hazard mitigation. These decisions, like those linked to response, can benefit from the rapid provision of information from modern seismic networks.

HAZUS loss estimates, based on monitored ground motion data (ShakeMaps), can expedite local, state, and federal disaster declaration processes by providing a potential basis for the preliminary damage assessment required under the provisions of the Stafford Act for the activation of state and federal resources. These estimates also facilitate a more efficient damage assessment of critical facilities (e.g., hospital damage and availability), help determine the resources needed to care for people in temporary housing, and provide the necessary information to mobilize state and federal disaster recovery programs. The 2001 Nisqually earthquake is a pioneering example of this process. State Governor Gary Locke received an initial HAZUS-based damage estimate 90 minutes after the earthquake, declared a state of emergency, and used the HAZUS estimates in a request for federal disaster assistance the next day.

Although regarded as a response activity, the disaster declaration process has major implications for an efficient recovery. If the costs associated with response and recovery from an earthquake disaster are judged to exceed the capacity of local government to pay without outside aid, the Stafford Act provides a mechanism for appeals to successively higher levels of government for financial and material assistance, up to and including a presidential declaration. At times, this process has been hampered by divergent estimates of community impacts and disagreements over resource needs. With the development of modern seismic networks—in combination with the loss estimation capabilities of HAZUS—this process can become more efficient and relatively free of conflict. HAZUS provides the rapid, reasonably accurate, and objective estimate of total dollar losses from an earthquake that is needed for a preliminary damage assessment. This assessment is an important component both in the decision to issue a disaster declaration and, if declared at the national level, in determining an appropriation from Congress for recovery.

Damage assessment involves the inspection, detailed description, and estimation of repair costs on a structure-by-structure basis following a damaging earthquake. Although HAZUS estimates losses on a regional rather than site-specific basis, the combined data from ShakeMap and HAZUS can identify census tracts in which buildings of a particular construction type may have sustained damage and thereby facilitate a prioritization of the assessment process. Additional information available from HAZUS, including average income and ethnicity, can alert recovery planners to the need for translators or information that would facilitate the administration of other assistance programs.

Many local, state, private nonprofit, and federal programs are available to assist individuals and organizations during recovery from a major disaster. They include loans and grants that help victims repair homes and businesses and pay mortgage or rent. These programs may also provide crisis counseling; disaster-related unemployment assistance; medical, dental, or funeral expenses; temporary housing; and other expenses. The contribution of seismic monitoring data to the efficient administration of these programs can be demonstrated by the use of monitoring data from the Northridge earthquake for recovery actions (Box 7.1).

RECENT RESPONSE EXPERIENCES

As noted above, the potential benefits of enhanced seismic monitoring are currently being fully realized only in the small area encompassing urban Los Angeles and San Francisco. Other areas, despite high or moderate levels of seismic hazard, will not gain such benefits without a concerted effort to enhance the seismic networks in these regions. Recent

BOX 7.1
Use of Monitoring Data for Recovery Assistance

The Federal Disaster Housing Assistance Program is available to renters and homeowners to cover the costs of alternative housing if disaster-related damages render a primary dwelling uninhabitable. To qualify, an applicant must complete an application, indicate that the dwelling is not habitable due to earthquake damage, and usually wait 2-5 weeks for an inspection to verify that the damage is severe enough to warrant an alternative housing arrangement. After the Northridge earthquake, the pre-grant inspection was waived. Instead, a zip code-based seismic intensity map, using source data from the Southern California Seismic Network, was used to determine grant eligibility. If an applicant's residence was in one of the 66 zip codes that corresponded to estimated Modified Mercalli intensities of VIII, IX, or X, the applicant was sent a check, reducing the delay between application and receipt of funds from weeks to a few days. A total of 49,000 checks—totaling $138 million in grants—were distributed. Verification inspections revealed that more than 90 percent of those receiving checks were eligible based on established program criteria (Goltz, 1996).

This rather bold experiment is instructive in several respects. First, it demonstrates that seismic monitoring data have applications in the recovery as well as the emergency response phase of an earthquake disaster. Second, it indicates a willingness by state and federal program administrators to use monitored data to streamline and expedite the delivery of program benefits, thus reducing the anxiety and possible suffering of disaster victims. Finally, the technologies associated with seismic monitoring and loss estimation have advanced significantly in the decade since the Northridge earthquake and can now provide data that are more refined, accurate, and reliable for application to program needs.

earthquakes outside California's two major urban areas provide illustrations of economic losses due to inadequate seismic network information.

In the aftermath of the Nisqually earthquake of February 28, 2001, the American Red Cross deployed disaster response workers to Seattle based on the magnitude and location of the earthquake. This limited information indicated that the event impacted urban Seattle and, at *Mw* 6.8 (similar to the 1994 Northridge earthquake), was expected to have caused casualties and damage. Although new seismic instruments were being installed in the region under the ANSS program, these instruments were not yet networked; thus, the immediate construction of a ShakeMap was not possible. At a depth of 52 km, the earthquake failed to generate the serious impacts anticipated by the Red Cross and resulted in an overdeployment, costing the organization an estimated $250,000.

On December 9, 2003, an *Mw* 4.5 earthquake occurred 15 miles southeast of Columbia, Virginia (see Figure 7.1), an event that was widely felt in Virginia, Maryland, and parts of North Carolina. The earthquake caused intensity VI shaking in central Virginia, prompting the evacuation of several office buildings and a jail in Richmond. Although the shaking was felt by many, earthquakes of this magnitude are unlikely to cause damage in engineered structures and the evacuations were unnecessary, resulting in disruption and loss of productivity. Enhanced seismic monitoring, including the placement of accelerometers in buildings and near infrastructure, expedites the immediate post-earthquake safety evaluation of structures and can be used either to avoid needless evacuations in smaller earthquakes or to expedite evacuations in larger events where the safety of occupants is jeopardized.

Although timely information was provided by the California Integrated Seismic Network following the recent *Mw* 6.5 San Simeon earthquake of December 22, 2003, this information proved to be inadequate for an effective emergency response. Because the initial ShakeMap was based on data from the only two seismic instruments available, the map failed to reflect the strong motion that caused damage in Paso Robles—an error that additional seismic instrumentation coverage would have prevented. Responding on the basis of this first iteration of ShakeMap, the California Transportation Department (CalTrans) deployed bridge and highway inspectors to Highway 1, rather than Route 101 where damage was much more likely because of the southeasterly direction of rupture propagation. HAZUS loss estimates, based on the earliest version of ShakeMap, underestimated damage.

As these recent experiences indicate, the ShakeMap and HAZUS outputs derived from modern seismic monitoring networks can provide significant benefits for emergency response and recovery. For those very few regions where sufficient monitoring is available, the benefits of such

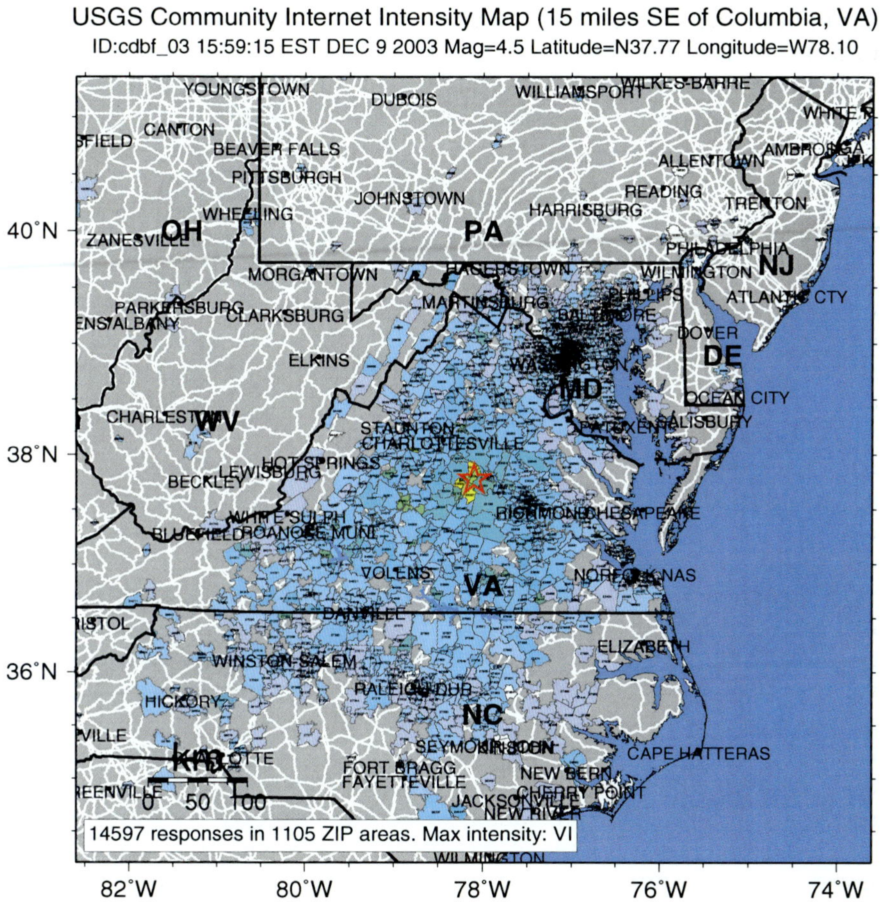

INTENSITY	I	II-III	IV	V	VI	VII	VIII	IX	X+
SHAKING	Not felt	Weak	Light	Moderate	Strong	Very strong	Severe	Violent	Extreme
DAMAGE	none	none	none	Very light	Light	Moderate	Moderate/Heavy	Heavy	Very Heavy

FIGURE 7.1 Community Internet Intensity Map for the *Mw* 4.5 earthquake near Columbia, Virginia on December 9, 2003.
SOURCE: USGS internet output. See http://pasadena.wr.usgs.gov/shake/ne/STORE/Xcdbf_03/ciim_display.html.

monitoring include rapid and accurate identification of the event, its location and magnitude, the extent of strong ground shaking, and estimates of damage and population impacts. This information enables rapid mobilization at levels appropriate to the emergency and facilitates rapid identification of areas requiring prioritized response.

SUMMARY

Seismic monitoring and the products that are derived from modern networks—including ShakeMap and HAZUS—offer significant benefits for emergency response and recovery. The benefits of enhanced seismic monitoring include rapid and accurate identification of the event, its location and magnitude, the extent of strong ground shaking, and estimates of damage and population impacts. This information expedites hazard identification, promotes rapid mobilization at levels appropriate to the emergency, and facilitates the rapid identification of buildings that are safe for continued occupation and those that must be evacuated. While it may be reasonable to surmise that—in the context of the $100 billion single-event estimate—improved (in particular, more targeted) emergency response would yield avoided costs in the range of tens of millions of dollars, there simply is insufficient information to provide a rigorous basis for quantitative estimate of potential benefits. Such estimates can be provided only by post-earthquake analysis of a region that has adequate seismic monitoring.

These are tangible benefits to the emergency management community and, ultimately, to residents of seismically active regions of the country. Although difficult to quantify, the ultimate benefits are lives saved, property spared, and human suffering and anxiety reduced. The trail that leads back from potential benefits to seismic monitoring is one of technological innovations that are directly linked to monitoring. Adequately monitored regions make robust ShakeMaps possible, by translating ground motion into the locations of potential damage. Armed with almost immediate information on such potential damage, emergency managers can first determine appropriate levels of response mobilization and then allocate those resources appropriately to search and rescue, fire suppression, transportation route recovery, and other response activities. How many lives are saved? How many fires with the potential to consume dozens of structures are suppressed before they are able to do so? How many injured people are spared avoidable suffering? Such benefits are difficult to quantify because there are so few regions where modern digital monitoring systems have been installed and because there have been no large damaging urban earthquakes in these areas since modern networks have been installed.

Large damaging urban earthquakes will occur again, and they will not necessarily occur in areas that are adequately monitored. However, if they do hit monitored areas, researchers will have an opportunity to assess quantitatively and qualitatively whether the use of ShakeMap, HAZUS loss estimates, and other products derived from monitoring made a significant difference in the timeliness and efficiency of emergency response. This assessment could be accomplished through careful content analysis of duty logs, after-action reports, hospital admission records, shelter records, and other documentation of response and recovery activities, comparing time frames and content with records of earthquakes of similar characteristics that occurred prior to the use of technologies based on seismic monitoring. Surveys could also be conducted to determine how these monitoring-based technologies were used and how effective they proved to be. Only over the last decade or so have seismic networks, because of significant advances in computing and software, developed capabilities that address the needs of emergency response and recovery. Partnerships between network operators and emergency managers in well-monitored regions are strong, new technologies are being integrated into response and recovery plans, and capabilities are waiting to be proven.

8

Integrating the Benefits— Conclusions and Recommendations

What does all the benefit information tell us? First, what is a relevant benchmark for examining whether the potential benefits from additional seismic monitoring can be justified given the anticipated costs of this investment? Annual costs for operating and maintaining the nation's *existing* seismic monitoring networks, including the present implementation of the federally operated network (the Advanced National Seismic System [ANSS]) and the plethora of state, university, and private networks, amount to approximately $31 million, although this is augmented by as-yet-unquantified state and local support. Costs for the seismic monitoring components of EarthScope—the permanent USArray additions to the national "backbone" as well as Plate Boundary Observatory (PBO) seismometers—amount to $4.9 million for hardware purchase and installation, with annual operations and maintenance (O&M) costs as yet unspecified (estimated at $1.5 million).

The expected capital expenditures for expansion and modernization of strong motion capabilities as part of the full ANSS proposal are slightly greater than $171 million, and expected annual operating costs—once the network is fully installed—are almost $47 million (in 1999 dollars) (USGS, 1999). In 2003 dollars, the equivalent amounts would be $189 million and $52 million, respectively. In total, capital expenditures for *improved* seismic monitoring (omitting costs for existing networks) amount to $194 million, with annual O&M costs of approximately $53 million. As is true for any project, these forecasts must be carefully hedged, and it is likely that actual annual expenditures will follow a path that reflects higher outlays in the early years—when construction and acquisition of the monitoring equipment occur—compared with later years when most of the costs will be for operations and maintenance. Similarly projections of benefits are likely to increase over time as practitioners learn to make the best use of data generated by improved seismic monitoring.

Because it is difficult to make plausible forecasts of the costs and benefits over the relevant future time horizon, we can proceed by considering a prototypical year. First, based on a presumed 10-year life for the new equipment, the annual costs would be the annualized equivalent of $194 million plus the annual $53 million annual O&M costs. This annual combined cost comes out to $96 million.[1] Second, what amounts of benefits are needed to justify these expenditures? The Federal Emergency Management Agency (FEMA, 2001a) reported that the expected annualized building-related earthquake losses to U.S. society are $4.4 billion. These figures include the repair and replacement costs for structural and nonstructural components, building content loss, business inventory loss, and direct business interruption loss. Based on the Consumer Price Index, these estimates were converted to 2003 dollars to give an annualized earthquake loss of $5.6 billion. FEMA's studies are conservative in that they do not include losses to utility and transportation systems, the costs of loss of life and injury, or indirect business interruption costs. Although other studies have estimated earthquake losses to be higher (e.g., Gordon et al., 2004), FEMA's more conservative estimates can be used here to evaluate whether the potential benefits can justify the anticipated improved seismic monitoring costs. If they pass this test with FEMA's data, then the investment will be even more attractive when compared with estimates from other more comprehensive studies.

One way to proceed with this discussion of whether the anticipated economic benefits from improved seismic monitoring justify the cost to the nation is to ask how large a reduction in expected annual earthquake losses would be required to justify the investment in improved seismic monitoring. Using the FEMA estimates of annualized buildings and building-related earthquake losses of $5.6 billion, if an annual 2 percent reduction in losses resulted from mitigation measures based on improved seismic monitoring data—a seemingly achievable result in light of the broad range of potential benefits described in this report—then avoided losses would be greater than the maximum expected annual costs of investing in increased seismic monitoring. A brief recap of the benefits elaborated in earlier chapters suggests that the investment in improved

[1]The annual opportunity cost of $194 million (4 percent cost of capital at 10 years) is approximately $24 million and the annual (straight-line) depreciation cost is approximately $19 million.

seismic monitoring should easily meet the efficiency (positive net benefit) hurdle.

SUMMARY OF BENEFIT COMPONENTS

A broad range of potential benefits of improved seismic monitoring is described in Chapters 4-7. These can be summarized in three categories—benefits that flow from information that is available immediately following a single earthquake or swarm of earthquakes (e.g., levels of ground shaking, potential for coastal inundation by a tsunami, or warnings about an imminent volcanic eruption); intermediate- to long-term benefits that occur as society responds to the information provided by monitoring data; and knowledge benefits that will accrue as a result of an improved fundamental understanding of earthquake processes and the distribution of earthquake risk.

Short-Term Benefits

Benefits Associated with Earthquake Emergency Response. In those areas of the country with modern digital seismic networks and rapid communication of information to a central processing site where data are rapidly analyzed and disseminated, emergency managers have many useful tools at their disposal for responding to a damaging earthquake. The deployment of digital strong motion recording instruments in southern California, northern California, the Pacific Northwest, and the Salt Lake City area as part of the initial deployment of the ANSS, together with improvements in data processing and communications during the last decade, have facilitated the creation and Internet distribution of Geographic Information System (GIS) based maps that show the location of strong ground shaking (ShakeMaps) within minutes to tens of minutes after an event. In the critical short-term period immediately after an earthquake, these ShakeMaps are invaluable for creating an initial "snapshot" of the emergency by providing descriptions of the intensity of shaking, identifying which jurisdictions have been affected, and providing the basis for prioritizing response activities. Responders use information from instrumented buildings and infrastructure, when available, to support their determination of the degree of damage and functional capabilities. One immediate benefit is the quicker reuse of monitored buildings and infrastructure following an event, without the need to wait for inspections. ShakeMaps generated from seismic network data are also the primary input for loss estimates from HAZUS, which provide more detailed information including estimates of the number of casualties, the number of people displaced from their homes, and the approximate number who will require shelter. Accurate information from the monitoring networks will also, through HAZUS, assist in the recovery effort by identifying areas of greatest economic impact and the direct and indirect economic losses. The restricted distribution of these significant benefits leaves the majority of earthquake-prone areas with either no information or less than optimal information upon which to mobilize a response. A fully funded ANSS will put important earthquake response tools in the hands of emergency responders in all seismically vulnerable regions of the nation.

Benefits Associated with Tsunami Warnings. Tsunamis present a relatively infrequent, yet significant danger to coastal communities in California, Washington, Oregon, Alaska, Hawaii, and Puerto Rico. Seismic monitoring of large and great subduction zone earthquakes around the circum-Pacific and Caribbean region provides valuable public safety and warning information in advance of tsunami arrivals. Although great strides have been made over the last 50 years in tsunami detection and warning, 75 percent of all tsunami warnings issued since 1948 have been false alarms and did not require evacuation. Not only are these false alarm evacuations costly, they also erode the credibility of the emergency management tsunami warning system. The cost of failing to evacuate for a real event or incorrectly estimating the risk, however, can be much greater. Improvements in seismic monitoring can significantly increase the accuracy of the tsunami warnings and reduce the risk of false alarms and missed warnings.

Benefits Associated with Volcanic Eruptions. Nearly every recorded volcanic eruption has been preceded by an increase in earthquake activity beneath or near the volcano. For this reason, seismic monitoring has become one of the most useful tools for eruption forecasting and monitoring. The overall economic risk to aircraft from airborne volcanic ash is estimated to be about $70 million per year (Kite-Powell, 2001). Coordinated observations, using both land- and space-based data, are needed to evaluate volcanic threats in realtime. Seismic monitoring—coupled with satellite observations and ash-cloud transport models—enables the air transportation industry to reroute flights and avoid costly ash-cloud encounters.

Intermediate- to Long-Term Benefits

Intermediate- and long-term benefits result from society's reaction to seismic monitoring information in a strategic manner beyond the immediacy of an emergency response. The incremental benefits in this time frame principally reflect additional loss avoidance activities, beginning with property damage and running the course of all loss categories. Such loss avoidance would result from either information gained from a single event or the accumulation of monitoring information over time. This accumulated knowledge will result in an improved approach to the design and construction of buildings and infrastructure, implementation of appropriate mitigation of existing structures, and the refinement of building polices and regulations.

Benefits from Improved Seismic Zonation. The pattern of damage caused by earthquakes often has a highly irregular distribution, with concentrations of damage in some locations and relatively little damage in others. It is typical for the ground motion level to vary by a factor of two, or sometimes much more, between different locations that are equally close to the earthquake source. The capability to reliably predict the pattern of ground motion amplification in urban areas (seismic zonation), and thus identify locations that are especially vulnerable as well as ones that are not, has the potential to significantly reduce the aggregate cost of seismic mitigation by allowing it to be better focused on highly vulnerable areas. In the long term, this will lead to a reduction in earthquake losses and will guide rational urban development. However, the development of this capability is contingent on the deployment of dense arrays of strong motion recording instruments in urban regions, as is planned for ANSS.

Benefits from Improved Earthquake Recurrence Models. Much of our knowledge of the location and characteristics of active faults that can generate potentially damaging ground shaking and other hazards, and the frequency of occurrence of potentially damaging earthquakes, is derived from a combination of geologic mapping and seismic monitoring. Knowledge about the source characteristics of large and small earthquakes, derived from recordings from seismic networks and other data, is used to identify and characterize the seismic potential of earthquake sources throughout the United States. Improved seismic monitoring will reduce uncertainty in earthquake recurrence and earthquake source characteristics throughout the United States, providing a more reliable basis for the ground motion maps used as the basis for building codes.

Benefits from Improved Ground Motion Prediction Models. The ground motion prediction models used in current building codes are highly uncertain because of the sparse data set of strong ground motion recordings on which these models are based. This high level of uncertainty gives rise to large predicted ground motions at low annual probability levels (i.e., 1 in 2,475 chance of occurrence, corresponding to a 2 percent chance of occurrence in the next 50 years) that forms the basis of the FEMA National Earthquake Hazards Reduction Program (NEHRP) seismic provisions. In most regions of the United States, the ground motion design levels on which these building codes are based are larger than any ground motions that have ever been recorded in those regions. Deployment of ANSS would, over time, provide data that would reduce the uncertainty in ground motion models and hence result in refined design levels for use in building codes. It could potentially also confirm indications from large overseas earthquakes that our current ground motion models, based on extrapolation of data from smaller-magnitude events at larger distances, may overpredict the ground motions close to large earthquakes.

Benefits from Improved Loss Estimation Models. Improved seismic monitoring will improve the accuracy of the data underpinning loss estimation models and reduce the uncertainty associated with model outputs (see Chapter 5), leading to a number of beneficial economic impacts:

- Improved model credibility will increase public knowledge, confidence, and understanding of seismic risk. Pre-event planning will be improved and more focused as the range of potential outcomes is reduced.
- Building code and land-use requirements will improve. Reduced uncertainty will allow more effective correlation between seismic risk and building code and land-use regulations. In some instances this will lead to more rigorous standards and in others to less rigorous standards.
- Reduced uncertainty in the output of the loss estimation models will increase the amount of coverage that insurance and reinsurance companies are able to provide, and since reduced uncertainty will reduce risk, the cost of earthquake coverage should drop. Improved insurance take-up rates should shift more of the cost of disasters from grants and disaster relief payments (funded by the taxpaying public at large) to insurance recoveries (financed through premiums paid by property owners and tenants).
- Reduced uncertainty—and increased confidence in loss estimation models—will enable local, state, and federal decision-makers to better monitor the growth of seismic risk in the nation. Information about new and rehabilitated buildings and infrastructure, coupled with improved seismic hazard maps, will allow policy-makers to track incremental improvements in seismic safety through earthquake mitigation programs.

Benefits Provided by Performance-Based Engineering. The seismic hazard that exists across the nation is now quantified in scientifically defensible seismic hazard maps, and the inherent uncertainty in the methodology is used as a basis for the conservatism that these maps embody. Engineering technology has advanced to the extent that new buildings can be designed and existing structures rehabilitated to selected levels of safety. This technology—performance-based engineering—uses a combination of seismic hazard and structural vulnerability to produce facilities having predictable performance. Conservatism is built into the design process to compensate for the expected uncertainty through the use of probabilistically based ground motion estimates, subjective limits on the use of various structural systems and commonly used materials of construction, and numerous additional safety factors. The prospective benefits from performance-based engineering are based on implementation of improvements in the design and rehabilitation processes compared with those currently used throughout the nation. A percentage of the buildings that are instrumented are expected to suffer negligible damage during severe earthquakes. This will constitute a proof test, and consequently the cost of rehabilitation will be avoided. Similarly, a percentage of buildings that are instrumented and damaged will require fewer repairs because their specific strengths and vulnerabilities will be better understood. The availability of site- and building-specific monitoring data will improve the ability to predict the performance of buildings under design-level earthquakes, leading to a reduction in the number of buildings that require strengthening and a reduction in the amount of strengthening necessary in those still found to be deficient. In addition, the improved seismic monitoring information is expected to yield improved hazard maps and refined design techniques that will yield more seismically efficient, and consequently less costly, designs. Special emphasis should be placed on instrumenting publicly owned buildings, especially federal buildings, to ensure continuity of maintenance of the instruments, open access to information about structural design and construction history, timely access to the monitoring records, and avoidance of liability issues that may concern private building owners.

The discussion of the benefits of performance-based engineering (Chapter 6) makes the case that improved seismic design, expected to become available as a result of improved seismic monitoring, will generate prospective annual benefits of more than $140 million. Benefit calculations were based on an estimated valuation of the built environment now exposed to significant seismic risk of about $9 trillion and growing by about $500 billion per year. Since, with the exception of the West Coast states, most construction was not based on seismic design considerations, it is conservative to assume that 10 percent of the buildings need significant seismic strengthening, and strengthening costs will amount to between 10 and 150 percent of building value. For new construction, the added costs of seismic design are in the range of 1-10 percent of building value.

Knowledge Benefits

The third category of benefits is the accretion of knowledge. The accumulation of information from improved seismic monitoring potentially leads to a more complete understanding of spatial and temporal physical processes associated with faulting and other sources of seismic activity. The accumulated record of weak and strong motion information could ultimately lead to some type of earthquake prediction capability.

Potential for Benefits from Earthquake Prediction. At present there is no operational capability for short-term earthquake prediction, and it is unclear whether such a capability will ever be developed. However, several approaches to earthquake prediction are currently being tested, and deployment of the ANSS would provide valuable information for the development and testing of earthquake prediction methods. The ability to reliably predict future damaging earthquakes would completely transform our current approach to earthquake loss reduction and risk management. The current approach is based not on knowing the times and locations of future damaging earthquakes, but on assessments of their long-term frequency of occurrence. From a longer-term perspective, if we could predict or forecast damaging earthquakes that will occur in the United States over the next several years to decades, we would know where to focus mitigation activities and where such activities could be deferred. This would result in a large increase in the benefit-cost ratio of mitigation activities, because the benefits of mitigation in the face of otherwise certain losses would be greatly enhanced, and all of the resources available for mitigation could be devoted to locations where losses would otherwise certainly occur. If we could predict the damaging earthquakes that will occur over the intermediate term (the next several months to several years), there might not be time to complete much structural mitigation, but there could be a focus on preparedness activities that might potentially reduce both direct and indirect losses as well as deaths and injuries. Short-term prediction of damaging earthquakes (the next several hours to days) would permit a wide range of preparedness activities. These might include evacuation of hazardous locations, suspension of plans for non-emergency surgery at hospitals in favor of preparation to handle injuries, and securing lifelines and vital business and government data management operations. These actions could potentially reduce both direct and indirect losses as well as deaths and injuries.

BENEFIT INTEGRATION

Seismic monitoring plays a significant role in decision-making and benefit estimation by reducing the uncertainty associated with risk assessments; aiding the process of risk perception; and enabling individuals, groups, and organizations to make more informed choices. In addition, improved data on the likelihood and consequences of earthquakes derived from seismic monitoring facilitates the identification of avoided costs and losses through the development of more meaningful risk management strategies ranging from emergency preparedness and mitigation to programs for aiding the recovery process following an earthquake. A variety of policy instruments (e.g., economic incentives, insurance, building codes) and regulations can be used to reduce future earthquake damage and loss of lives while providing financial relief after a disaster.

The capability to reliably predict the pattern of ground motion amplification in urban areas, and thus identify locations that are especially vulnerable as well as ones that are not, has the potential to reduce earthquake losses significantly and guide rational urban development. The development of this capability is contingent upon the deployment of dense arrays of strong motion recording instruments in urban regions, as planned for the ANSS. There are very few recordings close to the large-magnitude earthquakes that control the design of structures in the western United States, and there are very few strong motion recordings of even small earthquakes in the eastern and central United States. Until there is adequate monitoring of strong ground motion in the United States, earthquake engineering practice will continue to be subject to very great uncertainty in the ground shaking levels that are appropriate for design, with the concomitant economic effects of potential underdesign (resulting in needlessly high damage levels) or overdesign (resulting in needless construction costs).

There are clearly a large number of significant benefits that can be associated with improved seismic monitoring. Do they amount to at least a 2 percent reduction of estimated earthquake losses? A precise answer, involving a re-estimation and updating of FEMA's HAZUS annualized loss study, may be unnecessary in light of the benefits discussions elaborated in Chapters 4-7. The dollar estimates for just some of the potential annual benefits from improved performance-based engineering add up to more than $140 million, an amount considerably higher than the range of estimated annual costs. This corroborates the idea that the relatively low annual costs of ANSS are such that the efficiency hurdle would be met. Clearly a full calculation of prospective benefits across all the benefit classes identified in this analysis, including lives saved, would reach a much higher prospective benefits total. The results of benefit-cost analyses, by themselves, are simply an input into the complex policy-making process. However, even with a set of conservative assumptions, implementation of the ANSS will yield prospective benefits that substantially exceed expected costs and therefore meet economic efficiency tests. Although an improved base level of research and information—much of which can be gained only after substantial improvements in the nation's seismic monitoring capabilities—is required before a rigorous and fully quantified estimate of potential benefits can be made, the analysis undertaken here indicates that on an annual basis the dollar costs for improved seismic monitoring are in the tens of millions and the potential dollar benefits are in the hundreds of millions.

RECOMMENDATIONS

The combination of the earthquake hazard together with the vulnerability of the built and human environment creates earthquake risk. The earthquake risk in the United States is growing at an alarming rate, even in the face of the remarkable advances in earthquake science and engineering (EERI, 2003). This phenomenon is the direct result of unprecedented growth and prosperity and the lack of focused, nationally applied, public policy that would cause the available design and rehabilitation techniques to be properly and universally applied. Earthquakes continue to cause an unacceptable level of damage, in terms of lives lost, property destroyed, and service interruption. Reduced levels of uncertainty and increased confidence in loss estimation models will enable local, state, and federal decision-makers to monitor the long-term growth of seismic risk in the United States. Information about the seismic performance of new and rehabilitated buildings and infrastructure, coupled with improved seismic hazard maps, will allow policy-makers to track incremental improvements in seismic safety through implementation of earthquake mitigation programs.

The potential benefits from improved seismic monitoring are quite varied. An important role of seismic information is to improve the accuracy (i.e., reduce the uncertainty) of building damage predictions and loss estimates, as the basis for more effective loss avoidance regulations, as well as enabling more effective emergency preparedness activities and improved earthquake forecasting capabilities. Each earthquake provides a unique opportunity to learn. Improved monitoring of future earthquakes will lead to a more complete understanding of geophysical processes, more effective hazard mitigation strategies, and improved emergency response and recovery.

As with all projects designed to reduce losses from natural disasters, the ANSS is expected to provide benefits in the form of avoided losses. Consequently, the costs of earthquake damage to the nation—without mitigation measures based on data and information provided by ANSS—must form the benchmark against which the prospective benefits are assessed. Losses or costs associated with earthquakes fall into five major categories—direct physical damage (to buildings and infrastructure), induced physical damage (including fires, floods, hazardous material release, etc.), human impacts (death and injuries), costs of response and recovery (including first-responder and building inspection costs, etc.), and business interruption and other economic (social and environmental costs, etc.) losses. The most recent estimate of annual earthquake losses in the United States by the FEMA was $5.6 billion per year for buildings and building-related costs (after adjustment to 2003 dollars; building-related direct economic losses include repair and replacement costs for structural and nonstructural components, building content loss, business inventory loss, and direct business interruption losses), with a single significant earthquake potentially causing losses greater than $100 billion. Although concentrated on the West Coast, the risk of significant earthquake loss applies to many areas of the country. In recognition of the magnitude and extent of potential losses:

> **The United States should rank arresting the future growth of seismic risk and reducing the nation's current seismic risk as highly as other critical national programs that need persistent long-term attention, and it should make the necessary investment to achieve these goals.**

Our understanding of the nature of earthquake hazards in the United States—the distribution, frequency, and severity of damaging ground shaking—is based on past damaging earthquakes as well as on the tens of thousands of small earthquakes that occur throughout the nation each year. Improved seismic monitoring networks will provide the basis for better characterization of this seismicity, so that the ground motion prediction models that underpin building codes and earthquake engineering design—the basis for safeguarding life and property—can more accurately reflect the complex nature of the hazard. In addition, any potential for the future prediction of damaging earthquakes will rely in part on seismic monitoring data.

Estimates of the extent of likely damage and the socioeconomic consequences of earthquakes are based on loss estimation models, which combine seismic hazard and vulnerability models with inventories of the built environment. Loss estimation models are contained in commercial software packages and in publicly available models, the most widely known and used in the latter category being the HAZUS model. HAZUS is a standardized, nationally applicable, multihazard loss estimation methodology for estimating the impacts of disasters for the purposes of risk mitigation, emergency preparedness, and disaster recovery. All loss estimation models share a common structure; they are based on an estimate of the severity of the earthquake hazard, coupled with engineering estimates of the damage and loss to the infrastructure inventory in a particular region. In some loss model applications, the frequency of the hazard is also considered in order to provide the end-user with probabilistic loss estimates rather than scenario loss estimates. Output from the models typically includes the amount of expected damage to the built environment, economic costs of that damage (including business interruption costs), and estimates of injuries and deaths. Loss estimation models are used by insurers and reinsurers, government agencies, private businesses, the engineering community, and others. Improved seismic monitoring will enhance the accuracy of the data underpinning loss estimation models, and reduce the uncertainty associated with model outputs.

The benefits from improved loss estimation model outputs include increased public knowledge, confidence, and understanding of seismic risk; better correlation between seismic risk and building code and land-use regulations; more efficient use of insurance to offset losses from disasters; and more accurate determination of the nature and growth of seismic risk in the nation. In addition, information about new and rehabilitated buildings and infrastructure, coupled with improved seismic hazard maps, will allow policy-makers to track incremental improvements in seismic safety through earthquake mitigation programs.

Improved earthquake hazard assessments combined with more accurate loss estimation models—both dependent on improved seismic monitoring—offers significant benefits for emergency response and recovery. These benefits include rapid and accurate identification of the event, its location and magnitude, the extent of strong ground shaking, and estimates of damage and population impacts. This information expedites hazard identification, promotes rapid mobilization at levels appropriate to the emergency, and facilitates the rapid identification of buildings that are safe for continued occupation and those that must be evacuated. These are tangible benefits to the emergency management community and, ultimately, to residents of seismically active regions of the country. Although difficult to quantify, the ultimate benefits are lives saved, property spared, and reduced human suffering.

The integration of HAZUS loss estimation capabilities and U.S. Geological Survey (USGS) earthquake hazard information should be continued to track the growth of seismic risk in the United States, thereby further reducing the uncertainty associated with seismic risk.

The guidelines, standards, and codes available to earthquake engineers for the design of new structures and the rehabilitation of existing structures hold promise for protecting lives and the built environment against the largest expected earthquakes. However, perceptions that the up-front cost of mitigating the risk of earthquake damage is too high, combined with skepticism concerning the likelihood of earthquake occurrence—particularly in areas that have not experienced damage in historical time—leaves the country in a state of increasing seismic risk with the rapid expansion of the built environment. In order to make significant advances in arresting the growth of seismic risk, new analysis and design techniques are needed to better accommodate the expected ground motion. Current engineering design guidelines are mostly based on field observations that result in generalized and conservative procedures for controlling damage. The recent excellent performance of buildings where motion has been recorded provides a reasonable expectation that new techniques can be developed that will reduce the cost of seismic safety to more affordable levels. Seismic monitoring records hold the key to understanding how the built environment responds to significant earthquakes, and improved records offer the potential for fine-tuning the design process so that seismic safety requirements are adequately—but not excessively—met.

Determining the value of information has always been a challenge—it is not a tangible commodity and its benefits are often very subtle. Additional specific limitations apply to seismic monitoring information where the positive result of the information is avoided loss (e.g., in retrospective studies, it is difficult to isolate the contribution of seismic monitoring from other factors that influence the reduction in earthquake losses). Nevertheless, public policy decisions generally have to be made despite such limitations. The relative gains provided to society by improved monitoring information can be measured by the economic value of reduced decision uncertainty, assessed by comparing actions to be taken to manage the risks with and without improved monitoring.

It is possible, by using a series of assumptions, to determine a ballpark figure for earthquake losses that could be avoided by using improved seismic monitoring information as the basis for implementing improved performance-based earthquake engineering design. These assumptions relate to the value of the built environment within the United States, the cost of seismic rehabilitation and the number of existing buildings that need strengthening, and the annual expected loss from earthquakes compared with reduced losses when *higher* seismic design standards based on information from improved monitoring are applied. These calculations indicate a total loss avoided of more than $140 million per year, based on an estimate of reduced earthquake losses together with estimates of savings in construction costs that would accrue from the implementation of performance-based engineering design in those regions where improved seismic monitoring indicates that seismic design standards can be *reduced*.

Although it is possible to compile qualitative descriptions of the existing uncertainties and the potential economic benefits of improved seismic monitoring, existing research and information are insufficient to provide a full quantitative assessment of such benefits. In effect, a certain level of seismic monitoring information—to be provided by the monitoring proposed for the ANSS—will be required before rigorous quantitative determination of the benefits can be made. The extent of the assumptions required to make the ballpark calculations for performance-based engineering design emphasizes the need for additional quantitative information before more precise estimates of the economic benefits of seismic monitoring can be determined.

After every damaging earthquake in the United States, data gathering and applied research should be sponsored—as a collaborative activity among the National Earthquake Hazards Reduction Program (NEHRP) agencies—to document how seismic monitoring information reduced uncertainty and provided economic benefits in both the long and the short term. Comprehensive reports should be published within one year after the event for short-term benefits, and within 10 years after the event for intermediate- and long-term benefits.

The relatively modest funding required to significantly improve seismic monitoring should be viewed in light of the potential for reducing the cost of constructing new facilities, strengthening existing structures to achieve proper performance, and avoiding losses after major damaging events. The approximately $200 million investment required for improved seismic monitoring infrastructure should be considered from the perspective of the more than $800 billion invested annually in building construction, the $17.5 trillion value of existing buildings in the United States, and the possibility of a $100 billion plus loss from a single, major earthquake in an heavily populated urban environment.

After assessing the considerable range of potential economic benefits from improved seismic monitoring that will be provided by full implementation of the ANSS, the committee concludes:

Full deployment of the ANSS offers the potential to substantially reduce earthquake losses and their consequences by providing critical information for land-use planning, building design, insurance, warnings, and emergency preparedness and response. In the committee's judgment, the potential benefits far exceed the costs—annualized buildings and building-related earthquake losses alone are estimated to be about $5.6 billion, whereas the annualized cost of the improved seismic monitoring is about $96 million, less than 2 percent of

the estimated losses. It is reasonable to conclude that mitigation actions—based on improved information and the consequent reduction of uncertainty—would yield benefits amounting to several times the cost of improved seismic monitoring.

References

Abrahamson, N.A., and K.M. Shedlock, 1997. Overview of Ground Motion Attenuation Models. *Seismological Research Letters,* 68: 9-23.

Allen, C.R., 1976. Responsibilities in earthquake prediction. *Bulletin of the Seismological Society of America,* 66: 2069-2074.

ATC (Applied Technology Council), 1991. *Seismic Vulnerability and Impact of Disruption of Lifelines in the Conterminous United States.* Redwood City, California, Applied Technology Council Report ATC-25; 440 pp.

ATC (Applied Technology Council), 2001. *Database on the Performance of Structures Near Strong-Motion Recordings: 1994 Northridge, California, Earthquake.* Redwood City, California, Applied Technology Council Report ATC-38; 260 pp.

Bakun, W.H., and A.G. Lindh, 1985. The Parkfield, California earthquake prediction experiment. *Science,* 229: 619-624.

Bakun, W.H., and V.T McEvilly, 1984. Recurrence models and Parkfield, California earthquakes. *Journal of Geophysical Research,* 89: 3051-3058.

Bakun, W.H., F.G. Fischer, E.G. Jensen, and J. Van Schaak, 1994. An Early Warning System for Aftershocks. *Bulletin of the Seismological Society of America,* 84: 359-365.

Bernknopf, R., D. Brookshire, and M. Thayer, 1990. Earthquake and Volcano Hazard Notices: An Economic Evaluation of Changes in Risk Perception. *Journal of Environmental Economics and Management,* 18: 35-49.

Bernknopf, R.L., D.S. Brookshire, D.R. Soller, M.J. McKee, J.F. Sutter, J.C. Matti, and R.H. Campbell, 1993. *Societal Value of Geologic Maps.* Denver, Colorado, U.S. Geological Survey, Circular 1111; 53 pp.

Bernknopf, R.L., D.S. Brookshire, M. McKee, and D.R. Soller, 1997. Estimating the Social Value of Geologic Map Information: A Regulatory Application. *Journal of Environmental Economics and Management,* 32: 204-218.

BSSC (Building Seismic Safety Council of the National Institute of Building Sciences), 2004. NEHRP Recommended Provisions for Seismic Regulations for New Buildings and Other Structures (FEMA 450), 2003 Edition. Prepared by the BSSC for the Federal Emergency Management Agency, Washington, D.C. Available on-line at http://www.bssconline.org/; accessed October 2005.

CNSS (Council of the National Seismic System), 1998. Network Information Summaries from CNSS and other Networks. Available on-line at http://www.anss.org/CNSS/NETS/; accessed July 2004.

Cochrane, H.C., 1997. Forecasting the Economic Impact of a Midwest Earthquake. Pp. 223-248 *in* B.G. Jones (ed.), *Economic Consequences of Earthquakes: Preparing for the Unexpected.* Buffalo, New York, National Center for Earthquake Engineering Research.

EarthScope, 2002. *EarthScope: Scientific Targets for the World's Largest Observatory Pointed at the Solid Earth.* Workshop Report, Snowbird, Utah, October 10-12, 2001; 56 pp.

EERI (Earthquake Engineering Research Institute), 2003. *Securing Society against Catastrophic Earthquake Losses: A Research and Outreach Plan in Earthquake Engineering.* Oakland, California; 62 pp.

EFEC (EarthScope Facilities Executive Committee), 2003. *EarthScope: Acquisition, Construction, Integration and Facility Management.* National Science Foundation Proposal; Major Research Equipment and Facilities Construction Account. Washington, D.C.; 174 pp.

Eguchi, R.T., J.D. Goltz, C.E. Taylor, S.E. Chang, P.J. Flores, L.A. Johnson, H.A. Seligson, and N.C. Blais, 1998. Direct Economic Losses in the Northridge Earthquake: A Three-Year Post-Event Perspective. *Earthquake Spectra*, 14: 245-264.

ELCON (Electricity Consumers Resource Council), 2004. The Economic Impacts of the August 2003 Blackout. Available on-line at http://www.elcon.org/Documents/EconomicImpactsOfAugust2003Blackout.pdf; accessed May 2005; 10 pp.

Ellsworth, W.L., 1990. Earthquake History, 1769-1989. Pp. 152-187 *in:* R.E. Wallace (ed.), *The San Andreas fault system, California.* Denver, Colorado, U.S. Geological Survey, Professional Paper 1515.

Ellsworth, W.L., A.G. Lindh, W.H. Prescott, and D.G. Herd, 1981. The 1906 San Francisco Earthquake and the Seismic Cycle. Pp. 126-140 *in:* D.W. Simpson and P.G. Richards (eds.), *Earthquake Prediction—An International Review.* Maurice Ewing Series 4, Washington, D.C., American Geophysical Union.

Enarson, E., and B.H. Morrow (eds.), 1998. *The Gendered Terrain of Disaster: Through Women's Eyes.* Westport, Connecticut, Praeger; 288 pp.

Fedotov, S.A., 1965. Regularities in the distribution of strong earthquakes in Kamchatka, the Kuril Islands, and northeast Japan. *Institut Fiziki Zemli Trudy, Akademia Nauk SSSR*, 36: 66-94.

Feinstein, D., 2001. *S. 424. A bill to provide incentives to encourage private sector efforts to reduce earthquake losses, to establish a national disaster mitigation program, and for other purposes; to the Committee on Finance.* Pp. S1754-S1760 *in:* Congressional Record—Senate, March 1, 2001, 147(26). Washington D.C., U.S. Government Printing Office.

FEMA (Federal Emergency Management Agency), 1997. *Report on Costs and Benefits of Natural Hazard Mitigation.* FEMA 294; 50 pp.

FEMA (Federal Emergency Management Agency), 2001a. *HAZUS 99 Estimated Annualized Earthquake Losses for the United States.* Mitigation Directorate, Washington, D.C., FEMA 366; 33 pp.

FEMA (Federal Emergency Management Agency), 2001b. *The 2000 NEHRP Recommended Provisions for Seismic Regulations for New Buildings and Other Structures, 2000 Edition. Part 1: Provisions.* Washington, D.C., FEMA 368; 374 pp.

FEMA (Federal Emergency Management Agency), 2001c. *The 2000 NEHRP Recommended Provisions for Seismic Regulations for New Buildings and Other Structures, 2000 Edition. Part 1: Commentary.* Washington, D.C., FEMA 369; 446 pp.

FEMA (Federal Emergency Management Agency), 2004. *HAZUS-MH Software, Version 1.0.* Released January 2004.

Field, E.H., 2001. Earthquake Ground Motion Amplification in Southern California. U.S. Geological Survey, Open-File Report 01-164. Available on-line at http://pubs.usgs.gov/of/2001/of01-164/; accessed May 2005.

FMI Corporation, 2004. *The 2003-2004 U.S. Markets Construction Update.* Raleigh, North Carolina, FMI Corporation.

Frankel, A.D., D.L. Carver, and R.A. Williams, 2002. Nonlinear and linear site response and basin effects in Seattle for the M 6.8 Nisqually, Washington earthquake. *Bulletin of the Seismological Society of America,* 92: 2090-2109.

Freeman, A.M., III, 2003. *The Measurement of Environmental and Resource Values: Theory and Methods, Second Edition.* Washington, D.C., Resources for the Future; 496 pp.

Ganderton, P.T., 2005. 'Benefit Cost Analysis' of Disaster Mitigation: Application as a Policy and Decision-Making Tool. Pp. 445-465 *in:* C.E. Haque, (ed.), *Mitigation of Natural Hazards and Disasters: International Perspectives,* Springer.

Goltz, J.D., 1996. Use of Loss Estimates by Government Agencies in the Northridge Earthquake for Response and Recovery. *Earthquake Spectra,* 12: 441-445.

Gordon, P., H. Richardson, and B. Davis, 1998. Transport-Related Impacts of the Northridge Earthquake. *Journal of Transportation and Statistics,* 1: 22-36.

Gordon, P., J.E. Moore II, H.W. Richardson, M. Shinozuka, and S. Cho, 2004. Earthquake Disaster Mitigation for Urban Transportation Systems: An Integrated Methodology that Builds on the Kobe and Northridge Experiences. Pp. 205-232 *in:* Y. Okuyama and S.E. Chang (eds.), *Modeling Spatial and Economic Impacts of Disasters,* Berlin, Springer Verlag.

Graves, R.W., A. Pitarka, and P.G. Somerville, 1998. Ground motion amplification in the Santa Monica area: effects of shallow basin edge structure. *Bulletin of the Seismological Society of America,* 88: 1224-1242.

Grossi, P., and H. Kunreuther (eds.), 2005. *Catastrophe Modeling: A New Approach to Managing Risk.* New York, Springer Verlag; 252 pp.

Guffanti, M., and E.K. Miller, 2002. Reducing the threat to aviation from airborne volcanic ash. Presentation to 55th Annual International Air Safety Seminar, 4-7 November 2002, Dublin.

Harris, R.A., 1998. Forecasts of the 1989 Loma Prieta California earthquake. *Bulletin of the Seismological Society of America,* 88: 898-916.

Hartzell, S., E. Cranswick, A. Frankel, D. Carver, and M. Meremonte, 1997. Variability of Site Response in the Los Angeles Urban Area. *Bulletin of the Seismological Society of America,* 87: 1377-1400.

Heinz Center (The H. John Heinz III Center for Science, Economics and the Environment), 2000. *The Hidden Costs of Coastal Hazards.* Washington, D.C., Island Press; 220 pp.

Heinz, J.A., and C.D. Poland, 2001. Correlating measured ground motion with observed damage. Pp. 110-130 *in:* J. Uzarski and C. Arnold (eds.), Chi-Chi, Taiwan, Earthquake of September 21, 1999: Reconnaissance Report. *Earthquake Spectra, Supplement A to Vol. 17,* Oakland, California, Earthquake Engineering Research Institute (EERI), Publication No. 2001-02.

Hill, D.P., J.P. Eaton, and L.M. Jones, 1990. Seismicity 1980-86. Pp. 115-151 *in:* R.E. Wallace (ed.), *The San Andreas fault system, California.* Denver, Colorado, U.S. Geological Survey, Professional Paper 1515.

Hogarth, R., and H. Kunreuther, 1989. Risk, Ambiguity and Insurance. *Journal of Risk and Uncertainty,* 2: 5-35.

Huber, O., R. Wider, and O.W. Huber, 1997. Active Information Search and Complete Information Presentation in Naturalistic Risky Decision Tasks. *Acta Psychologica,* 95: 15-29.

Iboshi, P. I., 1996, Tsunami Alert Economic Loss Estimation—memo to State of Hawaii, Civil Defense, May 22, 1996.

ICBO (International Conference of Building Officials), 1973. *Uniform Building Code, 1973 Edition*. California, Whittier; 131 pp.

ICC (International Code Council, Inc.), 2000. *International Building Code*. Building Officials and Code Administrators International, Inc., Country Club Hills, Illinois; International Conference of Building Officials, Whittier, California; and Southern Building Code Congress International, Inc., Birmingham, Alabama.

Jackson, D.D., and Y.Y. Kagan, 1991. Seismic gap hypothesis: Ten years after. *Journal of Geophysical Research*, 96: 21419-21431.

Jackson, D.D., and Y.Y. Kagan, 1993. Reply to Comment, Seismic gap hypothesis: Ten years after. *Journal of Geophysical Research*, 99: 9917-9920.

Jaume, S., and L.R. Sykes, 1999. Evolving towards a critical point: A review of accelerating seismic moment/energy release prior to large and great earthquakes. *Pure and Applied Geophysics*, 155: 279-306.

Jones, L.M., and P.A. Reasenberg, 1989. Some Facts about Aftershocks to Large Earthquakes in California. U.S. Geological Survey, Open-file Report 96-266. Available on-line at http://geopubs.wr.usgs.gov/open-file/of96-266/; accessed October 2005.

Keilis-Borok, V.I., 1996. Intermediate-term earthquake prediction. *Proceedings of the National Academy of Sciences USA*, 93: 3748-3755.

Kelleher, J., L.R. Sykes, and J. Oliver, 1973. Possible criteria for predicting earthquake locations and their application to major plate boundaries of the Pacific and the Caribbean. *Journal of Geophysical Research*, 78: 2547-2585.

Ketchum, M, V. Chang, and T. Shantz, 2004. *Influence of Design Ground Motion Level on Highway Bridge Costs*. Pacific Earthquake Engineering Research Center, Report 6D01. Available on-line at http://peer.berkeley.edu/lifelines/LL-CEC/reports/final_reports/6D01/6D01-FR.pdf; accessed May 2005; 59 pp.

Kite-Powell, H.L., 2001. Benefits of NPOESS for Commercial Aviation—Volcanic Ash Avoidance. Woods Hole Oceanographic Institute (WHOI). Available on-line at http://www.economics.noaa.gov/library/documents/benefits_of_observing_systems/benefits-npoess-commercial_aviation.pdf; accessed May, 2005.

Kunreuther, H., R. Hogarth, and J. Meszaros, 1993. Insurer Ambiguity and Market Failure. *Journal of Risk and Uncertainty*, 7: 71-88.

Kunreuther, H., N. Novemsky, and D. Kahneman, 2001. Making Low Probabilities Useful. *Journal of Risk and Uncertainty*, 23: 103-120.

Kunreuther, H., R. Meyer, and C. Van den Bulte, 2004. *Risk Analysis for Extreme Events: Economic Incentives for Reducing Future Losses*. NIST Technical Report GCR 04-871. Washington, D.C., National Institute of Standards and Technology; 93 pp.

Leith, W., 2005. *Progress developing the Advanced National Seismic System*. Address to Seismological Society of America Annual Meeting, Lake Tahoe, April 27, 2005.

Lindh, A.G., 1983. Preliminary assessment of long-term probabilities for large earthquakes along selected fault segments of the San Andreas fault system in California. U.S. Geological Survey, Open File Report 83-63; 15 pp.

Lindh, A.G., 2003. The nature of earthquake prediction. *Seismological Research Letters*, 74: 723-725.

Luken, R., F. Johnson, and V. Kibler, 1992. Benefits and Costs of Pulp Paper Effluent Controls Under the Clean Water Act. *Water Resources Research*, 28: 665-74.

McNutt, S.R., 2002, Volcano seismology and monitoring for eruptions. Pp. 383-406 *in:* W.H.K. Lee, H. Kanamori, P.C. Jennings, and C. Kisslinger (eds.), *International Handbook of Earthquake and Engineering Seismology, Part A*. Amsterdam, Academic Press.

Mileti, D., 1999. *Disasters by Design: A Reassessment of Natural Hazards in the United States*. Washington D.C., Joseph Henry Press; 351 pp.

Mileti, D.S., C. Fitzpatrick, and B.C. Farhar, 1992. Lessons from the Parkfield Earthquake Prediction. *Environment*, 34: 16-39.

Mitchell, R.C., and R.T. Carson, 1989. *Using Surveys to Value Public Goods: The Contingent Valuation Method*. Washington, D.C., Resources for the Future; 488 pp.

Mogi, K., 1968. Sequential occurrences of recent great earthquakes. *Journal of Physics of the Earth*, 16: 30-36.

Mogi, K., 1969. Some features of recent seismic activity in and near Japan (2), Activity before and after great earthquakes. *Bulletin of the Earthquake Research Institute, Tokyo University*, 47: 395-417.

Munich Re Group, 2000. *Topics 2000: Natural Catastrophes – The Current Position*. Available on-line at http://www.munichre.com/publications/302-02354_en.pdf; accessed March, 2004.

Munich Re Group, 2002. *Enclosure 1: The ten largest natural catastrophes of 2001*. Available on-line at http://www.munichre.com/Assets/pdf/press/PM_2001_12_28_anhang1_e.pdf; accessed May, 2005.

NIBS-FEMA (National Institute for Building Sciences-Federal Emergency Management Agency), 2002, *HAZUS®99 Earthquake Loss Estimation Methodology, Service Release 2 (SR2) Technical Manual*, Developed by the Federal Emergency Management Agency through agreements with the National Institute of Building Sciences, Washington, D.C.

Nishenko, S.P., 1991. Circum-Pacific seismic potential: 1989-1999. *Pure and Applied Geophysics*, 135: 169-259.

Nishenko, S.P., and L.R. Sykes, 1993. Comment on Seismic gap hypothesis: Ten years after. *Journal of Geophysical Research*, 98: 9909-9916.

NRC (National Research Council), 1973. *Strong-Motion Engineering Seismology: The Key to Understanding and Reducing the Damaging Effects of Earthquakes*. Washington, D.C., National Academy of Sciences; 17 pp.

NRC (National Research Council), 1980. *U.S. Earthquake Observatories: Recommendations for a New National Network*. Washington D.C., National Academy Press; 122 pp.

NRC (National Research Council), 1989. *Estimating Losses from Future Earthquakes*. Washington D.C., National Academy Press; 248 pp.

NRC (National Research Council), 1990. *Assessing the Nation's Earthquakes: The Health and Future of Regional Seismograph Networks*. Washington D.C., National Academy Press; 80 pp.

NRC (National Research Council), 1992. *The Economic Consequences of a Catastrophic Earthquake: Proceedings of a Forum*. Washington D.C., National Academy Press; 196 pp.

NRC (National Research Council), 1996. *Review of Recommendations for Probabilistic Seismic Hazard Analysis: Guidance on Uncertainty and Use of Experts*. Washington D.C., National Academy Press; 72 pp.

NRC (National Research Council), 1999. *The Impacts of Natural Disasters: A Framework for Loss Estimation*. Washington D.C., National Academy Press; 68 pp.

NRC (National Research Council), 2003a. *Living on an active earth: perspectives on earthquake science*. Washington, D.C., The National Academies Press; 418 pp.

NRC (National Research Council), 2003b. *Preventing Earthquake Disasters: The Grand Challenge in Earthquake Engineering*. Washington, D.C., The National Academies Press; 172 pp.

NRC (National Research Council), 2003c. *The National Earthquake Hazards Reduction Program at Twenty-Five Years: Accomplishments and Challenges — Summary of a Workshop, February 20, 2003, Washington, D.C.*, The National Academies Press; 21 pp.

OES (California Governor's Office of Emergency Services), 1990. *Short-Term Earthquake Response and Advisory Plan*. Sacramento, California.

Olson, R.A., and R.S. Olson, 2001. Socioeconomic Reverberations of Earthquake Prediction: Snapshot in Time, Peru 1979-1981. *Natural Hazards Review*, 2: 124-131.

Palm, R., 1998. Demand for Disaster Insurance: Residential Coverage. Pp. 51-66 *in:* H. Kunreuther and R.J. Roth, Sr., (eds.), *Paying the Price: The Status and Role of Insurance Against Natural Disasters in the United States.* Washington, D.C, Joseph Henry Press.

Peacock, W.G., B.H. Morrow, and H. Gladwin (eds.), 1997. *Hurricane Andrew: Ethnicity, Gender and the Sociology of Disasters.* London, Routledge; 277 pp.

Peek-Asa, C., J.F. Kraus, L.B. Bourque, D. Vimalachandra, J. Yu, and J. Abrams, 1998. Fatal and Hospitalized Injuries Resulting from the 1994 Northridge Earthquake. *International Journal of Epidemiology*, 27: 459-465.

Pitarka, A., K. Irikura, T. Iwata, and H. Sekiguchi, 1998. Three-dimensional simulation of the near-fault ground motion for the 1995 Hyogo-ken Nanbu (Kobe), Japan, earthquake. *Bulletin of the Seismological Society of America*, 88: 428-440.

Porter, K.A., J.L. Beck, and R.V. Shaikhutdinov, 2002. Sensitivity of Building Loss Estimates to Major Uncertain Variables. *Earthquake Spectra,* 18: 719-743.

Porter, K., K. Shoaf, and H. Seligson, in press. Technical Note: Value of Injuries in the Northridge Earthquake. *Earthquake Spectra.*

Randall, A., 1983. The Problem of Market Failure. *Natural Resources Journal,* 23: 131-148.

Reasenberg, P.A., and L.M. Jones, 1994. Earthquake Aftershocks: Update. *Science,* 265: 1251-1252.

Reid, H.F., 1910. *The Mechanics of the Earthquake, The California Earthquake of April 18, 1906. Report of the State Investigation Commission, Vol.2.* Washington, D.C., Carnegie Institution of Washington.

Rose, A., 2004. Economic Principles, Issues, and Research Priorities in Natural Hazard Loss Estimation. Pp. 13-36 *in:* Y. Okuyama and S. Chang (eds.), *Modeling the Special Economic Impacts of Natural Hazards.* Heidelberg, Springer.

Rose, A., and D. Lim, 2002. Business Interruption Losses from Natural Hazards: Conceptual and Methodological Issues in the Case of the Northridge Earthquake. *Environmental Hazards: Human and Policy Dimensions,* 4: 1-14.

Rose, A., and W. Miernyk, 1989. Input-Output Analysis: The First Fifty Years. *Economic Systems Research,* 1: 229-271.

Schiff, A., 1999. *Guide to the Improved Earthquake Performance of Electric Power Systems.* ASCE Manuals and Reports on Engineering Practice No. 96. American Society of Civil Engineers, Reston, Virginia; 341 pp.

Scholz, C.H., 1985. The Black Mountain asperity: Seismic hazard of the southern San Francisco Peninsula, California. *Geophysical Research Letters,* 12: 717-719.

Scholz, C.H., 1990. *The Mechanics of Earthquakes and Faulting.* Cambridge University Press; 439 pp.

Schwartz, D. P., and K.J. Coppersmith, 1984. Fault Behavior and Characteristic Earthquakes from the Wasatch and San Andreas Faults. *Journal of Geophysical Research,* 89: 5681-5698.

SEAOC (Structural Engineers Association of California), 1995. *Vision 2000: Performance-Based Seismic Engineering of Buildings.* Sacramento, California, Structural Engineers Association of California; 115 pp.

Seed, H.B., and I.M. Idriss, 1982. *Ground Motions and Soil Liquefaction During Earthquakes.* Earthquake Engineering Research Institute, Engineering Monograph on Earthquake Criteria, Structural Design, and Strong Motion Records 5, El Cerrito, California; 134 pp.

Seligson, H.A., and K.I. Shoaf, 2003. Human Impacts of Earthquakes. Pp. 28-1 to 28-29 *in:* W.-F. Chen and C. Scawthorn (eds.), *Earthquake Engineering Handbook.* Boca Raton, Florida, CRC Press.

Shah, H.C., and D. Rosenbaum, 1996. Earthquake Risk Shakes Mortgage Industry. *Secondary Mortgage Markets,* 13: 12-19.

Shinozuka, M., A. Rose, and R. Eguchi (eds.), 1998. *Engineering and Socioeconomic Impacts of Earthquakes.* Buffalo, New York; Multidisciplinary Center for Earthquake Engineering Research, Publication MCEER-98-MN02.

Shoaf, K., and L.B. Bourque, 2001. *Survey of Potential Early Warning System Users.* Center for Public Health and Disasters, University of California, Los Angeles.

Shoven, J.B., and J. Whalley, 1992. *Applying General Equilibrium.* Cambridge, UK, Cambridge University Press; 309 pp.

Slovic, P., 2000. *The Perception of Risk.* London, UK, Earthscan; 473 pp.

Smith, V.K., and S. Pattanayak, 2002. Is Meta-Analysis a Noah's Ark for Non-Market Valuation? *Environmental and Resource Economics,* 22: 271-296.

Somerville, P.G., 2003. Magnitude scaling of the near fault rupture directivity pulse. *Physics of the Earth and Planetary Interiors,* 137: 201-212.

Spence, W., R.B. Herrmann, A.C. Johnston, and G. Reagor, 1993. *Responses to Iben Browning's Prediction of a 1990 New Madrid, Missouri, Earthquake.* Washington, D.C., U.S. Geological Survey, Circular 1083; 248 pp.

Stein, R.S., A.A. Barka, and J.H. Dieterich, 1997. Progressive failure on the North Anatolian fault since 1939 by earthquake stress triggering. *Geophysical Journal International,* 128: 594-604.

Sykes, L.R., and S.P. Nishenko, 1984. Probabilities of occurrence of large plate rupturing earthquakes for the San Andreas, San Jacinto, and Imperial Faults, California, 1983-2003. *Journal of Geophysical Research,* 89: 5905-5927.

Sykes, L.R., B.E. Shaw, and C.H. Scholz, 1999. Rethinking earthquake prediction. *Pure and Applied Geophysics,* 155: 207–232.

Tierney, K.J., 1997a. Business Impacts of the Northridge Earthquake. *Journal of Contingencies and Crisis Management,* 5: 87-97.

Tierney, K., 1997b. Impacts of Recent Disasters on Businesses: The 1993 Midwest Floods and the 1994 Northridge Earthquake. Pp. 189-222 *in:* B.G. Jones (ed.), *Economic Consequences of Earthquakes: Preparing for the Unexpected.* Buffalo, New York, National Center for Earthquake Engineering Research, NCEER Report No. NCEER-SP-0001.

Tierney, K.J., 2000. *Implementing a Seismic Computerized Alert System (SCAN) for Southern California: Lessons and Guidance from the Literature on Warning Response and Warning Systems.* Disaster Research Center, University of Delaware, Final Project Report, 45; 92 pp.

UNAVCO, 2005. Strainmeter Costs. Available on-line at http://www.unavco.org/pubs_reports/proposals/ pbo/budget/NorthernCaliforniaBudgetStrainmeter.pdf; accessed April, 2005.

URS Group, Inc., 2001. *A Report on the Cost-effectiveness of the TriNet Project.* Prepared for the Federal Emergency Management Agency, Washington, D.C., August 9, 2001, by URS Group, Inc., Gaithersburg, Maryland; 21 pp.

USGS (U.S. Geological Survey), 1999. A*n Assessment of Seismic Monitoring in the United States: Requirement for an Advanced National Seismic System.* Denver, Colorado, U.S. Geological Survey, Circular 1188; 55 pp.

USGS (U.S. Geological Survey), 2003a. *Is a powerful earthquake likely to strike in the next 30 years?* USGS Fact Sheet 039-03; Available on-line at http://pubs.usgs.gov/fs/2003/fs039-03/; accessed October 2005.

USGS (U.S. Geological Survey), 2003b. *ANSS—Reducing the Devastating Effects of Earthquakes.* USGS Fact Sheet 046-03; Available on-line at http://pubs.usgs.gov/fs/fs-046-03/fs-046-03.html; accessed October 2005.

USGS (U.S. Geological Survey), 2004. *Volcanic Ash—Danger to Aircraft in the North Pacific.* USGS Fact Sheet 030-97; Available on-line at http://pubs.usgs.gov/fs/fs030-97/; accessed October 2005.

USGS (U.S. Geological Survey), 2005. *National Seismic Hazard Mapping Project.* Available on-line at http://eqhazmaps.usgs.gov/; accessed January 2005.

Uzarski, J., and C. Arnold (eds.), 2001. Chi-Chi, Taiwan, Earthquake of September 21, 1999: Reconnaissance Report. *Earthquake Spectra, Supplement A to Vol. 17*, Oakland, California, Earthquake Engineering Research Institute (EERI), Publication No. 2001-02.

Webb, G.R., K.J. Tierney, and J.M. Dahlhamer, 2000. Business and Disasters: Empirical Patterns and unanswered questions. *Natural Hazards Review*, 1: 83-90.

WGCEP (Working Group on California Earthquake Probabilities), 1988. *Probabilities of Large Earthquakes Occurring in California on the San Andreas Fault.* U.S. Geological Survey, Open File Report 88-398.

WGCEP (Working Group on California Earthquake Probabilities), 1990. Probabilities of Large Earthquakes in the San Francisco Bay Region, California. U.S. Geological Survey, Circular 1053; 51 pp.

WGCEP (Working Group on California Earthquake Probabilities), 1995. Seismic hazards in Southern California: probable earthquakes, 1994 to 2024. *Bulletin of the Seismological Society of America*, 85: 379-439.

WGCEP (Working Group on California Earthquake Probabilities), 1999. *Earthquake Probabilities in the San Francisco Bay Region: 2000-2030—A Summary of Findings.* U.S. Geological Survey, Open-File Report 99-517; Available on-line at http://geopubs.wr.usgs.gov/open-file/of99-517/of99-517.pdf; accessed October 2005.

WGCEP (Working Group on California Earthquake Probabilities), 2003. *Earthquake Probabilities in the San Francisco Bay Region, 2002-2031.* U.S. Geological Survey, Open File Report 03-214; Available on-line at http://pubs.usgs.gov/of/2003/of03-214/OFR-03-214_FullText.pdf; accessed October 2005.

Wyss, M., and R.E. Habermann, 1979. Seismic quiescence precursory to a past and a future Kurile island earthquake. *Pure and Applied Geophysics*, 117: 1195-1211.

APPENDIXES

Appendix A

Excerpts from USGS Circular 1111

The following sections are reproduced from Bernknopf et al. (1993), "Societal Value of Geologic Maps."

GEOLOGIC INFORMATION AS A PUBLIC GOOD (pages 45-47)

Types of Information Discussion of the public-good attributes of information begins with the distinction between general and specific information. A frequently made argument (see Musgrave, 1959; Becker, 1975; Cohn, 1979) is that general information is a public good, while specific information is a private good. There is the presumption that general information possesses more of the characteristics of a public good, having a lack of exclusion possibilities (anyone can use the information) and a lack of congestion costs (there is no cost of competition in the use of the information).

As applied to geologic maps, general information is collected at a scale that would be valuable for a variety of regional planning decisions encompassing a set of choices for land uses such as highway route selection, waste repository siting, energy exploration, and development impacts.[1] Such information also would be available for a long period of time, given the slow rate of decay of its usefulness.

[1]We are aware that a map cannot be all things to all users. We generalize, despite Varnes' (1974) caution, because it is highly likely that a regional geologic map will have at least two users.

Specific information, on the other hand, is much more localized (for example, specific siting characteristics of interchanges along a single road right-of-way) and of much less use for further application. In essence, the collection of site-specific geologic information for determining the economic and environmental feasibility of siting a waste repository would be of little use in road planning unless the road is to be constructed in the same location as the proposed waste facility. As the information becomes more specific, the number of users becomes smaller. Thus, geologic information can be both general and specific information. Our concern is with geologic maps in the general information category. We note, however, that in compiling specific information, general information is often necessary to provide background data. In what follows, we discuss the economic concepts associated with the production of a public good in order to gain insight into the nature of regional geologic information. From this discussion we propose a series of testable hypotheses that can be examined empirically.

Public Goods Pure public goods have two key characteristics. First, it is impossible, or inefficient, to exclude anyone (nonrival in consumption) from consuming the good once it is produced.[2] The availability to other users is not diminished. Second, the production of the good is characterized by jointness of supply (Musgrave, 1959).[3] The extreme case of jointness of supply arises when the cost of the good is made up entirely of fixed costs. The key characteristics of a public good are discussed in more detail in this section, including a brief introduction to the "free-rider" problem.

Nonrival in Consumption Public goods are nonrival in consumption; that is, any one individual's consumption of the output does not reduce the consumption by others. Maps are available free to certain groups, readily available in certain repositories, and reproducible, so there is little reason to believe that any individual could be restricted from use. There is an obvious case of nonrival consumption for regional geologic maps.

A second aspect of the nonrival consumption argument is the ability to legally exclude others from making full use of information through the use of patents and copyrights. Such rules for exclusion are necessary for the private sector to have the appropriate incentive to produce map infor-

---

[2]For many types of public goods it may be technically impossible to exclude individuals from consumption of the good once it is produced. National defense is one such good, and, at a local level, police and fire protection also have this characteristic.

[3]Economic efficiency is achieved when the cost of production is equal to the market valuation of the good for the last unit produced. If the marginal cost is zero (all costs are fixed and would be incurred whether one unit or one hundred units were produced), economic efficiency requires that the good be "sold" at zero price—be made freely available.

mation that would be otherwise publicly provided information. Since individuals are able to obtain map information by not paying (a "free ride"), a private sector producer would not be able to recover the cost of production and would not provide the good.

Implementation of an exclusion scheme is difficult in the case of regional geologic map information because the range of potential users is large and dispersed. Effectively there is no way to implement a payment scheme. As a general rule, as information becomes more general, there is it a larger group of potential beneficiaries and there is less likelihood that exclusion is feasible. This point can be seen by comparing the possibilities of exclusion from use of a general theoretical development in seismology and earthquake prediction in California, relative to an engineering rehabilitation job on a building in Berkeley, Calif., which is used in a particular application. In the case of the general information, there may be a role for the government to produce such information.

Jointness of Supply The jointness of supply condition is fulfilled; that is, the per-unit production and distribution costs of regional geologic map information are near zero, while the per-unit costs of the information collection make up almost 100 percent of total per-unit cost. Regional geologic maps possess this characteristic, because the bulk of the costs of producing such maps are borne "up front,"[4] while the actual printing and distribution costs are relatively small. Because the printing costs are relatively low, the cost of serving an additional consumer also is small.[5] For example, the expected per-unit cost of information collection and synthesis for a 1:100,000-scale map covering Loudoun County, Virginia, is about $1.16 million, while the cost of production and distribution are about $8.44 per unit.[6] Therefore, excluding consumers once the good has been produced is inefficient.

"Free-Rider" Problem When the above two characteristics for a public good occur, in most cases, the private supply of this type of good such as a regional geologic map will yield inefficient market outcomes. Too little geologic information is produced, and a market failure ensues. This type of market failure is known as the "free-rider" problem. Free riders are individuals or groups who attempt to enjoy a good while not paying for it; it is impossible or inefficient to exclude them from the activity. The

---

[4]These costs comprise the data collection, organization, interpretation, coding, and other functions that precede the actual publication of the USGS geologic maps.

[5]See Matti and others (1988) for a presentation of relative cost figures.

[6]See Matti and others (1988, table 1) for the costs of producing a regional geologic map. Total costs of map compilation and publication are $21,100. The normal production run is 2,500 copies.

TABLE A-1 Payoffs to the provision of a public good [C=contribution; NC=no contribution]

		Person B	
		C	NC
Person A	C	4, 4	1, 5
	NC	5, 1	2, 2

nature of the free-rider problem may be illustrated as an application of the prisoner's dilemma (see Mueller, 1989, p. 8-17), summarized as follows:

Consider a simple economy with two persons. Each person begins with an endowment of $2 and each has two choices (strategies): to contribute to the provision of a public good or not to contribute. The public good is generated by summing the total contributions and multiplying this by 1.5 to reflect the consumer surplus (total area under the demand curve or total willingness to pay for a commodity) generated by the public good. The public good is enjoyed equally by both persons, and their payoffs are given by the value of the public good minus their contributions. The payoffs to these strategies are shown in Table A-1. For example, if both choose C (contribution), the total contribution is $4, the value of the public good is $6, and each person receives $4 as his net payoff. If both choose NC (no contribution), then they keep their endowment, so the payoff is $2 to each. If one contributes and the other does not, then the contributor receives $1 (his share of the public good is $3, and his contribution is $2), while the noncontributor receives $5 (his share of the public good is $3, and he keeps his endowment). The equilibrium outcome in this game is (NC, NC) with no public good being produced. This result is unfortunate, since the total payoff is clearly greater in the (C, C) outcome. The (NC, NC) outcome arises because NC is a dominant strategy for both of the persons in this economy. That is to say, it is not in either individual's interest to separately contribute to the public good since the payoff from this strategy is lower than from the strategy of not contributing.

The outcome for the general case of the pure public good is that private (voluntary) production will lead to suboptimal levels of production.[7] As a

[7]When we see some voluntary contributions in the "real world" it is usually the case that the good generates some private benefits (including the utility from donating) or that the good is not a pure public good. Some classes of public goods may be provided via the private sector, although not necessarily at efficient quantities. Cornes and Sandler (1986) and Bergstrom and others (1986) demonstrate these findings. In their setting, some individual is willing to privately provide an initial quantity of the public good because his marginal utility exceeds the cost of the good. For other individuals in the economy, the public good is viewed

result of this type of individual behavior, economics research has argued that the government should intervene to ensure proper provision of the good.[8]

THE ROLE OF GEOLOGIC MAP INFORMATION IN THE MAKING OF A DECISION (pages 21-23)

Reducing Information Uncertainty with Geologic Maps Chapter I describes geologic map information. For purposes of this chapter, we provide a brief description of the nature of geologic map information as it applies to regulatory decisionmaking. The geologic characteristics of a parcel of land are based on the type, structure, and engineering characteristics of the rock that are identified during the geologic mapping. This activity involves observation, sampling, analysis, and interpretation. The resolution of geologic map information generally increases with larger scale maps (more detailed) and with newer vintage maps.

Geologic maps can be interpreted to provide the basis for statistics that infer quantitative attributes about the geologic characteristics of a particular parcel of land. More detailed (larger scale) maps provide more accurate statistics (provide a likely reduction in the variance). In the demonstration of our applications, the available geologic map information (at different vintages and scales) for a region is used to produce derivative map information showing rock permeability and shear strength. A geologic map's vintage is an important consideration because it represents the status of interpretations, concepts, and models that continue to evolve over time. These geologic characteristics are considered important for our applications. Suppose a new geologic map of a region is produced at a

---

as an income transfer, and, where the public good is normal, the result is an increase in the willingness to pay for the good by these persons. The resulting response is analogous to a Cournot reaction and can be shown to lead to positive provision. The key initial assumption of this analysis is that at least one person's demand exceeds the cost of providing the first unit of the good. For most public goods this is not the case, and the outcome is that private provision will be at zero levels.

Where the cost of the public good is such that no one person is willing to supply *any* amount on his own, there is some debate as to whether the private market will supply a positive amount. Where use of the good may be prevented after it is made available (a club good), Bagnoli and McKee (1991) have shown that a "focal equilibrium" exists in which the good is supplied at efficient levels. Where the good is subject to ex post consumption by noncontributors, Isaac and others (1985) have shown that the good is generally undersupplied.

[8]The free rider problem must be overcome by compelling payments for public goods through a tax system with penalties for noncompliance.

scale that is more detailed than the existing geologic map, as is commonly the case. For instance, at the new map scale, perhaps faulting is delineated more clearly and boundaries between different rock types (contacts) are defined better than on the older, less detailed geologic map. Better delineation of the geologic attributes results because the newer and improved geologic map is based on more detailed, systematic observations. If this improved geologic map information leads to a different number of restricted parcels, then we have a measure of the value of the improved geologic map information. The net benefits of this improved information are the changes in the expected loss avoided, and the costs are the costs associated with acquiring the improved geologic map information.

Gathering geologic map information is a challenging process because of the scarcity of geologic outcrops, scarcity and expense of drill-hole data, and complexity of possible geologic interpretations. In general, the density of data required for the appropriate level of geologic resolution is a function of map scale. At any map scale, geologic information has an inherent uncertainty because observations cannot be made everywhere, and extrapolations must be made. Thus, the true value of a geologic attribute for a location is described by a probability distribution. The central point of the probability distribution is the expected value of the geologic attribute in a specific location. The variance of the attribute corresponds to the variability of the geologic attribute over the entire mapped area. This concept is illustrated in Figure A-1, in which the probability distribution of a geologic attribute is plotted for two different levels of information, d_1 and d_2, corresponding, for example, to existing and improved geologic information. *R* in Figure A-1 denotes a regulatory standard (threshold), and $\bar{g}_k$ denotes the expected value of the geologic attribute for a given locality or parcel of land. The 95-percent confidence interval (2σ) about the expected value is indicated for the distribution d_1 as $2\sigma_{VI}$. With this information, we fail to reject the hypothesis that the allowed standard, *R*, is met for this parcel of land, because it is within two standard deviations of the expected value, and H_0: $\bar{g}_k - 2\sigma_{v1} < R$.

With improved geologic map information, based on larger scale maps or newer field data, we have distribution d_2. The improved information is more precise and more detailed. As such, distribution d_2 has less uncertainty (in this case, a smaller standard deviation, $2\sigma_{V2}$ than d_1. The null hypothesis is rejected because the expected value of the geologic attribute minus two standard deviations is greater than the standard, *R*.

The rejection of a particular parcel by use of information from d_2 occurs because the information in d_2 is more precise, not because there is a bias in the original data, d_1. Additionally, note that the expected value of the geologic attribute has had the effect only of reducing the variance of the statistic.

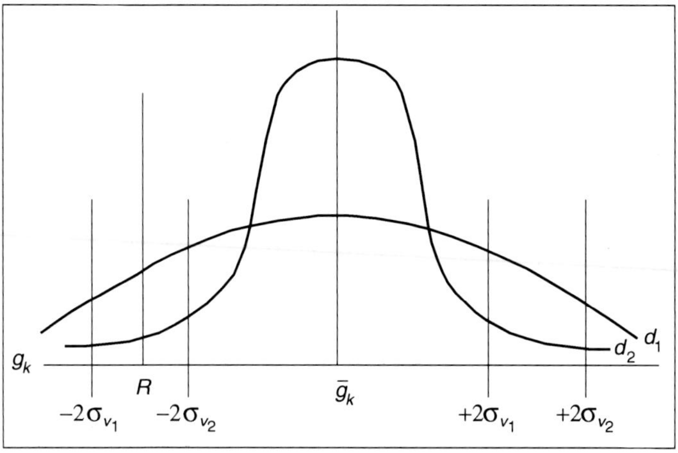

FIGURE A-1 The probability distributions, d_1 and d_2, of a geologic characteristic g_k, for two geologic maps of different vintages and different scales, v_1 and v_2, for the same area. R represents a regulatory standard. $\bar{g}_k$ denotes the expected value of the geologic attribute for a given locality or parcel of land. 2σ is the 95-percent confidence level about the expected value for the distributions.

The Inclusion of Geologic Map Information in Regulations Hypothetically, a regulatory agency could employ the following procedure to determine the number of parcels available for a given land use.[9] Each parcel is tested for whether the average value of the geologic attribute ($\bar{g}_k$), minus two standard deviations, is greater than or less than the regulatory standard. The rule in Equation A-1 is applied to each parcel.

$$\bar{g}_k - 2\sigma_v \begin{matrix} < \\ = \\ > \end{matrix} R \qquad \text{A-1}$$

where $\bar{g}_k$ is the average value of the geologic attribute in k, where k=l, . . ., K, σ_v is the standard deviation of g over the mapped area for a given vintage geologic map, v, and R is the regulatory standard.

[9]There are additional issues related to each parcel that go beyond the question of optimal information and the value of the information. For instance, there is the question of the optimal scale of the geologic map information for a specific land use application. In this study we are limited by the availability of the existing geologic maps. One could easily envision a study to address the question of the optimal scale for resolving a land use issue. In addition, there is the question, in a regional geologic mapping program, of prioritizing the schedule for creating geologic map information. Which geologic maps should be produced first? Again, this is beyond the scope of the study but is an important question in the overall issue regarding the "best" geologic mapping program.

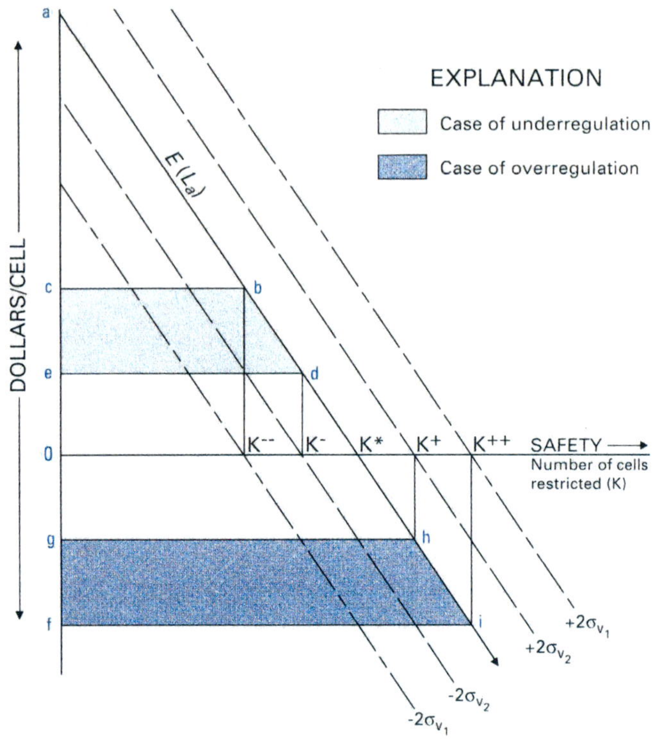

FIGURE A-2 Economic impact of a regulation based on geologic map information. $E(L_a)$ is the marginal expected loss avoided; K* is the optimal level of safety. See text for in-depth discussion.

The economic impact of a regulation based on the improved geologic map information can be seen in Figure A-2. This figure is a representation of the changes in the number of parcels restricted or mitigated when new, more detailed information becomes available. Safety (horizontal axis) is denoted as the fraction of land (parcels or cells) in the region that is rejected (require mitigation) for a particular land use type.[10] Losses avoided through implementation of regulations are measured in terms of a money metric (dollars) on the vertical axis. As the vulnerable, or "at risk," cells are eliminated, society's exposure to that risk is reduced. As the level of safety increases, there is an increase in expected losses avoided. The

[10]Henceforth, we shall refer to parcels of land as "k" rectangular cells in an equal-area grid to conform with the empirical work to be presented in this chapter.

change in expected losses avoided is shown in the figure as the marginal expected loss avoided, $E(L_a)$, for restricting each additional cell. The $E(L_a)$ curve represents the *net* marginal expected loss avoided (the cost of avoiding losses is constant), which is normalized on the figure by representing the $E(L_a)$ as deviations from zero. Therefore, an optimal level of safety is shown as the intersection of the $E(L_a)$ curve and the horizontal axis (net marginal expected loss avoided is zero) at the point labeled K^* in the figure.

There is uncertainty regarding the actual losses to be avoided by restricting cells, because there is uncertainty concerning the true state of the geology underlying a cell. We have indicated this uncertainty in Figure A-2 by the dashed lines above and below $E(L_a)$. We have indicated two levels of uncertainty, each of which is consistent with a different level of geologic map information (d_1 and d_2 in terms of Fig. A-1). In each case, the dashed lines enclose the 95-percent confidence interval.

The presence of information uncertainty leads to a tendency in the regulatory process to generate errors involving either underregulation or overregulation of land uses. Since the optimal level of safety is that which results in net marginal expected loss avoided being zero, either underregulation or overregulation must result in a welfare loss to society.

These losses may be shown by reference to Figure A-2. Consider the case in which the regulator sets the standards to restrict K^{--} cells. With K^{--} cells restricted, the social loss is given by the area bc0K*. This area is the amount of the potential consumer surplus that is foregone when the regulatory standard is set at K^{--} rather than the optimal level of K^*. Improved geologic map information results in the level of regulation being increased so that K^- cells are now restricted (this is the 95-percent confidence level with this improved information, d_2). The welfare loss is now given by the area de0K*. The value of the improved geologic map information is the gain in consumer surplus (the reduction in the welfare loss) shown as the area cbde in Figure A-2.

Consider now the case in which the regulator sets the standards at K^{++} (this is overregulation compared with the social optimal level of K^*). The welfare loss associated with this amount of overregulation in the area if0K*. With the improved information, the regulator reduces the number of cells restricted to K^+. The gain from this information is the area fihg.

REFERENCES

Bagnoli, M., and M. McKee, 1991. Voluntary contribution games: Efficient private provision of public goods. *Economic Inquiry*, 29: 351-366.

Becker, G., 1975. *Human capital.* 2nd edition. Chicago, Illinois, University of Chicago Press; 268 pp.

Bergstrom, T., L. Blume, and H. Varian, 1986. On the private provision of public goods. *Journal of Public Economics,* 29: 25-39.

Bernknopf, R.L., D.S. Brookshire, D.R. Soller, M.J. McKee, J.F. Sutter, J.C. Matti, and R.H. Campbell, 1993. *Societal Value of Geologic Maps.* Denver, Colorado, U.S. Geological Survey Circular 1111; 53 pp.

Cornes, R., and T. Sandler, 1986. The theory of externalities, public goods, and club goods. New York, Cambridge University Press; 451 pp.

Isaac, R.M., K. McCue, and C.R. Plott, 1985. Public good provision in an experimental environment. *Journal of Public Economics,* 26: 51-74.

Matti, J.C., R.L. Bernknopf, J.N. Van Driel, G.E. Ulrich, and J.S. Schindler, 1988. Photomechanical versus computer-based methods of preparing and disseminating geologic-map information: A comparison of costs and savings. U.S. Geological Survey administrative report, April 1988; 25 pp.

Mueller, D.C., 1989. *Public choice II.* New York, Cambridge University Press; 8-17.

Musgrave, R.A., 1959. *The theory of public finance.* New York, Macmillan; 628 pp.

Varnes, D.J., 1974. The Logic of Geological Maps, with Reference to Their Interpretation and Use for Engineering Purposes. Washington, D.C., U.S. Geological Survey, Professional Paper 837; 48 pp.

Appendix B

Committee and Staff Biographies

Chris D. Poland, Chair, is chairman, president, and chief executive officer of Degenkolb Engineers, a structural engineering firm specializing in earthquake engineering. Mr. Poland serves on a number of federal, state, and regional committees devoted to improving seismic safety and developing better codes and guidelines, and participates in various research activities sponsored by the National Institute of Standards and Technology and the Federal Emergency Management Agency. He is immediate past president of the Earthquake Engineering Research Institute.

James Ament recently retired from his position as vice president of operations of State Farm Fire and Casualty Company, the property insurance affiliate of the State Farm Group. Mr. Ament participated in policy development of enterprise risk management and the management of catastrophe insurance-related issues both within State Farm and with other interested groups.

David S. Brookshire is professor of economics and director of the Science Impact Laboratory for Policy and Economics at the University of New Mexico, specializing in public policy issues related to natural resource, natural hazard, and environmental economics. He received his B.A. from San Diego State University in 1970 and a Ph.D. in economics from the University of New Mexico in 1976.

James D. Goltz is an earthquake program manager in southern California for the statewide earthquake program of the California Governor's Office of Emergency Services, based at the Seismological Laboratory at the California Institute of Technology (Caltech) in Pasadena. He is responsible for promoting the application of new seismic information and technologies for improved emergency response and recovery. He has been engaged in earthquake preparedness, mitigation, response, and recovery in southern California for 25 years.

Peter Gordon is a professor in the Department of Economics and in the School of Policy, Planning, and Development at the University of Southern California. Dr. Gordon's research interests encompass urban and suburban sprawl and transportation networks, and the integration of transportation networks and regional economic models to estimate earthquake costs. Dr. Gordon has a B.A. from the University of California, Los Angeles; an M.A. from the University of Southern California; and a Ph.D. from the University of Pennsylvania.

Stephanie A. King is director of risk analysis at Weidlinger Associates, specializing in seismic hazard and risk analysis for regional and site-specific applications, and the use of these techniques to assess damage and loss due to natural and manmade hazards. She received her Ph.D. from Stanford University, where she developed automated computer techniques for probabilistic seismic hazard and risk assessment of large regions.

Howard Kunreuther is the Cecilia Yen Koo professor of decision sciences and public policy at the Wharton School, University of Pennsylvania, and serves as co-director of the Wharton Risk Management and Decision Processes Center. He has a long-standing interest in ways that society can better manage low-probability, high-consequence events as they relate to technological and natural hazards. He is a distinguished fellow of the Society for Risk Analysis and received the Society's Distinguished Achievement Award in 2001, and he is a recipient of the Elizur Wright Award for the publication that makes the most significant contribution to the literature of insurance.

Stuart Nishenko is senior seismologist in the Geosciences Department of the Pacific Gas and Electric Company (PG&E) in San Francisco, California, where he co-manages PG&E's Earthquake Risk Management Program and the University of California at Berkeley Pacific Earthquake Engineering Research Center Lifelines program on behalf of the California Energy Commission. He received his B.Sc. degree, magna cum laude, in geology from the City College of New York (1975) and M.S. (1978) and Ph.D. (1983) degrees in geophysics from Columbia University, Lamont-Doherty Geological Observatory.

Adam Z. Rose is a professor in the Department of Geography and former head of the Department of Energy, Environmental, and Mineral Economics at the Pennsylvania State University. Dr. Rose's research has been primarily in the areas of energy, environmental, and regional economics. He is the recipient of a Woodrow Wilson fellowship and the American Planning Association's Outstanding Program Planning Honor

Award. He received his Ph.D. in economics from Cornell University in 1974.

Hope A. Seligson is technical manager at ABS Consulting (formerly EQE International, Inc.), specializing in the area of seismic risk assessment, regional loss estimation, earthquake engineering, seismic hazard mapping, application of geographic information systems, software design and development, and emergency preparedness. She received her B.S. in civil engineering and M.S. in structural engineering, specializing in earthquake engineering, from Stanford University.

Paul G. Somerville is a strong motion seismologist at URS Group, Inc., where he has been involved in the development of seismological methods for specifying seismic design ground motions in earthquake engineering practice and has applied them in the design and analysis of major buildings, bridges, dams, and power generation facilities in many countries, especially the United States, Japan, and New Zealand. He has performed research for National Earthquake Hazards Reduction Program agencies, including the Federal Emergency Management Agency, National Science Foundation, and United States Geological Survey. He is involved in the development of seismic provisions of building codes and in the activities of professional research and practice organizations including the Earthquake Engineering Research Institute, Network for Earthquake Engineering Simulation, Pacific Earthquake Research Center, Southern California Earthquake Center, and the Seismological Society of America .

NATIONAL RESEARCH COUNCIL STAFF

David A. Feary is a senior program officer with the National Research Council's Board on Earth Sciences and Resources. His research activities have focused on the geological and geophysical evolution of continental margins, particularly the factors controlling carbonate deposition and reef development in different climatic regimes. He has B.Sc. and M.Sc. degrees from the University of Auckland and a Ph.D. from the Australian National University.

Appendix C

Acronyms and Abbreviations

AEL	Annual Earthquake Loss
AELR	Annual Earthquake Loss Ratio
ANSS	Advanced National Seismic System
BCA	Benefit-Cost Analysis
BSSC	Building Seismic Safety Council
CalTrans	California Transportation Department
CEA	California Earthquake Authority
CEPEC	California Earthquake Prediction Evaluation Council
CISN	California Integrated Seismic Network
COSMOS	Consortium of Organizations for Strong-Motion Observation Systems
CPI	Consumer Price Index
CSMIP	California Strong-Motion Instrumentation Program
CWB	Central Weather Bureau
DOE	Department of Energy
DOT	Department of Transportation
EERI	Earthquake Engineering Research Institute
EMS	Emergency Medical Services
EP	Exceedance Probability
FEMA	Federal Emergency Management Agency

GIS	Geographic Information System
GPS	Global Positioning System
HAZMAT	Hazardous Materials
IEEE	Institute of Electrical and Electronics Engineers
LADWP	Los Angeles Department of Water and Power
MCEER	Multidisciplinary Center for Earthquake Engineering Research
NEES	Network for Earthquake Engineering Simulation
NEHRP	National Earthquake Hazards Reduction Program
NEIC	National Earthquake Information Center
NIBS	National Institute for Building Sciences
NIST	National Institute of Standards and Technology
NOAA	National Oceanic and Atmospheric Administration
NRC	National Research Council
NSF	National Science Foundation
NSS	National Seismic System
OMB	Office of Management and Budget
PBO	Plate Boundary Observatory
PEER	Pacific Earthquake Research Center
PG&E	Pacific Gas and Electric
SAFOD	San Andreas Fault Observatory at Depth
SCEC	Southern California Earthquake Center
SEAOC	Structural Engineers Association of California
URM	Unreinforced masonry building
USGS	U.S. Geological Survey
USNRC	U.S. Nuclear Regulatory Commission
USNSN	U.S. National Seismic Network
WGCEP	Working Group on California Earthquake Probabilities
WWSSN	Worldwide Standardized Seismograph Network